FRONTIERS OF THE 21ST CENTURY: PRELUDE TO THE NEW MILLENNIUM

Edited by
Howard F. Didsbury, Jr.

Preface by
Howard F. Didsbury, Jr.

World Future Society
Bethesda, Maryland
U.S.A.

Editor: Howard F. Didsbury, Jr.

Editorial Review Committee: Deidre Banks, James J. Crider, Howard F. Didsbury, Jr. (Chairman) Charles H. Little, Theodore J. Maziarski, Frances Seagraves, Andrew A. Speakke, and Claudia Speakke

Staff Editors: Edward Cornish

Administrative Co-ordinator: Susan Echard

Editorial assistants: Gladys Archer, Robert Schley, Tanya Parwani-Jaimes

Cover: Lisa Mathias

Published by: World Future Society
7910 Woodmont Avenue
Suite 450
Bethesda, Maryland 20814
USA

International Standard Book Number: 0-930242-55-6

Printed in the United States of America

Contents

Preface
The Death of the Future in a Hedonistic Society
Howard F. Didsbury, Jr. vii

Introduction
Howard F. Didsbury, Jr. xii

Likely Developments with Universal Ramifications in the New Millennium
From Noosphere to Technosphere and Beyond
Joseph N. Pelton 3

Our Future as Postmodern Cyborgs
Chris Hables Gray 20

The Spectre of Emerging and Re-Emerging Infection Epidemics
Donald B. Louria, M.D. 41

Genetic Engineering: Our New Genesis
Clifton E. Anderson 59

Making Contact: The Most Important Event of the Coming Millennium?
Allen Tough 70

The Future of God
Robert B. Mellert 76

Do Universal Human Rights Imply the Future Development of a World Religion?
Charlotte Waterlow 83

Grow Old Along with Me. The Best(and) Worst Are Yet to Be
David Macarov 93

CompSpeak 2050: How Talking Computers Will Increase Oral Culture by the Mid-21st Century
William Crossman 106

The Global Impact of Information Technology: The Connected Versus the Unconnected
John W. McDonald 117

Prospects for Global Food Security in the 21st Century
Per Pinstrup-Andersen and Rajul Pandya-Lorch 121

Socio-Psychological Aspects of Information in a Democracy
Edward Wenk, Jr. 129

Utopias, Futures, and H. G. Well's Open Conspiracy
W. Warren Wagar 141

The Absolute, Urgent Need for Proper Earth Government
Robert Muller 148

The Next Thousand Years: The "Big Five" Engines of Economic Growth
Graham T.T. Molitor 155

The Creative Utilization of Human Capital
Creating Tomorrow's Dream Team
John Nance - Interview with Charles G. DeRidder 171

The Need for New Paradigms
John Diebold 175

Three Parallel Revolutions: Minding the Economic Laws of Knowledge
William E. Halal 206

Futurist Observations on a New Millennium
After the Party is Over: Futures Studies and the Millennium
Graham H. May 223

Creating and Sustaining Second-Generation Institutions of Foresight
Richard A. Slaughter 234

NOTE

Frontiers of the 21st Century: Prelude to the New Millennium was produced in conjunction with the World Future Society's Ninth General Assembly held in Washington, D.C., July 29-August 1, 1999. The Ninth General Assembly marks the 33rd anniversary of the founding of the Society. The assembly chairman was Kenneth W. Hunter. Susan Echard was assembly director. The assembly program director was Robert M. Schley.

The papers presented here were selected from a large number submitted to the Editorial Review Committee prior to the Ninth General Assembly. They are not necessarily papers presented by their authors at the assembly.

A number of distinguished papers whose subject matter did not lie within the limits of the volume could not be included.

As is our practice in general, footnotes and other academic paraphernalia have been minimized to avoid interruption in the flow of the author's ideas and insights.

PREFACE

THE DEATH OF THE FUTURE IN A HEDONISTIC SOCIETY

by

Howard F. Didsbury, Jr.

With the advent of what has been variously termed "post-industrial," "post-modern," "consumer," "information" or "high-tech" society, we have quickly grown accustomed to fantastic feats of technological wizardry, medical marvels, and spectacular scientific breakthroughs. The range of technological innovations and their further development and applications bewilder us. Though there may be some misgivings about the rapidity and nature of the resulting changes, there seems to be a quiet acquiescence to the seeming inevitable course of events. The transformations being brought about by the high-tech society and its dynamo of free market capitalism appears to have given rise to a cultural milieu which can best be described as a hedonistic society. In light of the exuberance occasioned by the demise of communist totalitarianism and the expectations associated with the apparent triumph of capitalism, perhaps we ought to reflect on the nature of this high-tech hedonistic society's characteristics insofar as it may become the norm for the highly developed parts of the world in the next century.

Just as privation and a day-to-day concern with mere survival tends to inhibit any genuine interest in the future—beyond the immediate demands of tomorrow—so likewise the advent of the hedonistic society, with its emphasis upon ease, comfort, immediate personal gratification and a life routine of continuous consumer distractions, tends to preclude any serious, sustained, long-range interest in the future. In both cases, any thought of positive action in the present to mold the future is discouraged, unrecognized or ignored!

The hedonistic society is captivated by the siren call of instantaneous gratification. While there is no iron-heeled repression, no thought police in such a society, as time passes very likely there is no serious or novel social thought to police. In this society there is no call of personal service or sacrifice for the general good but rather there is established an unassailable tyranny: of "presentism"—the concentration of interest on and the pursuit of everything for the here and the now.

In this society there is little or no concern for the past and little

---

Howard F. Didsbury, Jr. *is director of World Future Society special studies, professor emeritus of history, founder/director, Program for the Study of the Future, Washington, DC.*

thought of the future beyond what new commodity or excitement to-morrow may bring. Every now and then, almost by accident, there may be a sense of unease or disquiet when one thinks seriously, though fleetingly, of the future and what it may hold. The moment passes and the enchantment with the present quickly returns. In the sensate way of life the present is the watchword and yardstick of all importance and value.

In the place of citizens of a community or a nation as they have been understood in the conventional sense, the model high-tech society is peopled with constant, dedicated "consumers." In the society the good citizen is the tireless, voracious consumer whose constant consumption keeps the economic growth wheels turning and the society humming with activity. "...The declared objectives of increasing material prosperity and expanding industrial output can be maintained only by the vigilant cultivation of virtually insatiable appetites.[1] In passing it may be interesting to note the significance of an inordinate emphasis on a human being as a "consumer" of things in contrast to a consideration of the human being as an "enjoyer" of life. These two views are by no means the same. There is a famous temple and monastery in Kyoto, Japan, which has a water fountain at the entrance upon which is inscribed Chinese characters that read "I know I am wise for I know when enough is enough." This is an idea foreign to a sensate society.

The dynamism of the sensate society, it appears, is the mechanism to create endless wants. The seven capital sins (greed, avarice, envy, gluttony, sloth, lust, and pride) are all reevaluated to accommodate this way of life. Six of the seven are transformed into virtues in the hedonistic society. However, sloth remains a cardinal sin.[2] Possibly because the slothful in this society may not have the wherewithal to be a consumer.

The main technique employed to create catalogues of endless wants is applied psychology as exemplified in modern advertising. Over time it serves as a form of operant conditioning in the arousal and perpetuation of insatiable desires. In such a society everything can be made into a consumer item. In fact, even religion is "sold" as a personal consumer item. "I found faith and feel good."

> "...There is no need to enter the debate on the efficacy of commercial advertising in moulding people's tastes. Speaking only of the broad social repercussions of commercial advertising, one can hardly deny that it does appear to have succeeded wonderfully in one of its aims—that of making people discontented with what they already own. Indeed it is hard to imagine anything that would throw the American economy into greater disarray than a religious conversion that made most Americans perfectly contented with their material lot."[3]

In addition to this quest for the acquisition of things and for immediate personal gratification, there appears to be a desire to simplify everything. The cultural ideal could be stated as "Great rewards with little effort." Nowhere is this more evident than in the hedonistic society's attitude toward education. "...The ever-growing lack of hardship in the education of modern adolescents tends to make them less productive members of society..."[4] Decades ago the Nobel laureate Dennis Gabor observed that a permissive education in a permissive society was a prescription for carelessness, if not incompetence. The majority of problems which humanity faces today are largely the result of human ignorance, a lack of concern, or greed. The futurist as a tempered optimist believes that with knowledge, wisdom, and determination the problems can be solved or greatly alleviated. In contrast, in the sensate society, the easy and effortless solution to whatever the challenges may be, is the one sought. The hope seems to be that somebody somewhere somehow will be able to meet any challenges or difficulties that may occur.

A decade ago an editorial in *The Washington Post* (August 28, 1991) noted "After a decade of school reforms, rising budgets for education, reorganizations, and exhortation, the academic performance of American schoolchildren shows no improvement whatsoever [today we would probably say 'little']" In the spirit of the hedonistic society, "Typical American students don't get much out of high school because they don't put much into it. Adults are responsible for that. Most parents, most teachers and most politicians support the idea that it's all right for kids to drift through adolescence, waste time and, to the exclusion of all else, have fun." "The adults...created a climate in which there's not much appreciation of the deep rewards of competence, of developing real control over a subject, of learning to learn and often not much example of them among parents or teachers either." The course of least effort seems to be the one chosen by all concerned.

Tests, examinations, or any exercises designed to demonstrate results of effort expended or mastery of material are subjects of almost endless studies and debates. The ideas of challenge or performance are shunned everywhere but in the sports arena. Why this is so may not to be too difficult to fathom. First, sports are a form of entertainment; but also, in sports, it is not aspirations, intentions, or expectations that count—only results do!

Thanks to the Spirit of the Counter Culture which is still with us in many ways we have the "Word to the Wise"—if something bugs you—bug out! The idea of persistence—dedication and devotion to the attainment of a desirable goal for a better society or world—is now considered quite passé. If the goal or ideal cannot be achieved quickly—forget it. An appeal to history may be helpful especially when we realize that the overwhelming number of ideas we take for granted today as being the most sensible way of doing things were at one time considered "radical," "crackpot," impractical, or utopian.

By sheer persistence in working for their realization human beings brought about a great advance in the human condition.

The hedonistic society seems to be a society devoid of any serious interest in posterity. With its focus on the present and on the most efficient way of obtaining immediate results, any thought of the generations to come is ephemeral. It is this frame of mind which makes the threat of environmental degradation so frightening and makes diminished biological diversity on the planet a virtual certainty. With respect to elected officials, there is little need to overly concern oneself with future generations as posterity does not vote in the next election.

In keeping with the major concern of the "citizen" of the sensate society, the political agenda concentrates on personal "rights"with vague passing reference, if any, to concomitant responsibilities. On a superficial level there is considerable interest elicited with respect to personal perils and risks associated with environmental degradation. Frequently much clamor arises with demands for a pollution-free, wholesome physical and social environment, however, there is usually tenacious resistance to any programs that necessitate increased taxes or involve inconvenience. It is not a question of getting something for nothing. Rather the attitude resembles that of a spoiled child who wants everything for nothing. "The technological utopia in which every human whim and fancy finds complete and instant gratification is outside the realm of feasibility."[5]

The thesis which we presented at the outset states simply that a hedonistic society inhibits any serious concern about the future because it is a society which is fundamentally oriented toward individual ego satisfaction here and now. Its sense of the future all too easily becomes a vision of an exciting cornucopia of endless novel consumer items and continuing remarkable scientific and technological innovations. Missing in such a society is a genuine interest in how present action or inaction may affect present and future generations. Since the focus of this society is on the present, there is scant regard for speculations about a better society or world. When considered at all, speculation about human betterment (not human perfection) is usually dismissed with a contempt rivaling that of Karl Marx toward those he called "Utopian Socialists." For his part, Nobel laureate Ilya Prigogine declares " I am...afraid of a *lack* of utopias. I am afraid of the frying out of incentive. For example, if you think about politics for a moment, life becomes very uninteresting if incentives for conduct are limited strictly to economic exchanges. However, when we bring in the idea of nature, and visions of the natural world we would like to live in, or the idea of other civilizations, and the relationships we would like to have with them, 'politics' takes on a whole new meaning."[6] Decades ago, Alfred North Whitehead, one of the great philosophers of the 20th century, stressed the fact that there can be no great age without a great idea to inspire it. What will be the great idea to inspire the 21st century and set the tone for the next millennium?

NOTES

1. Mishan, E.J. "On Making the Future Safe for Mankind," *The Public Interest*, No. 24, Summer 1971, p. 56.
2. Observation made by Lewis Mumford in his discussion of the industrial era value system.
3. Mishan, E.J. p. 50.
4. Stent, Gunther S. *The Coming of the Golden Age: A View of the End of Progress*. Garden City, NY: The Natural History Press, p. 66, 1969.
5. Camilleri, Joseph A. "Human Prospects in a Changing World," *Civilization in Crisis*. Cambridge: Cambridge University Press, p, 220, 1976.
6. Prigogine, Ilya. "Beyond Being and Becoming" [An interview] in *New Perspectives Quarterly*, Vol. 9, No. 2, p. 28, Spring 1992.

INTRODUCTION

by

Howard F. Didsbury, Jr.

This pre-conference volume of the World Future Society's Ninth General Assembly consists of 20 papers, a rich and varied variety, that have been organized under three main headings:

Likely Developments with Universal Ramifications in the New Millennium

The Creative Utilization of Human Capital

Futurist Observations on a New Millennium

"From Noosphere to Technosphere and Beyond" by Professor Joseph N. Pelton the reader learns that "Today's Global Village is surging with intellectual growth which can be equated to a Telesphere or even an emerging Global Brain. This growth of information and communications (currently expanding at a rate of 300,000 times faster than our population) is changing our world and opening up new opportunities as well as key threats to our biosphere."

"There are at least four key challenges to the human species by the exponentially expanding telesphere:

- Seeking effective ways to manage the ever-accelerating global Niagara of 'mega-data', to preserve a livable biosphere, and to cope with the emerging 'Global Brain.'

- Finding ways to achieve humanizing 'Human-machine interfaces' and adjusting to a non-stop world where information systems are on a 168-hour work week.

- Exploring what might be the ultimate challenge for "intelligence in the universe which might be actually attempting to overcome entropy."

- Exploring the meaning of the ever-expanding Telesphere and what it means for the survival or even further evolution of the human species.

"The central assurances of modernism are no longer reassuring. Faith in inevitable progress has become insecurity about the future. A belief in the truths of science and the consistent utility of technological innovations has been supplanted not by a rejection of science and technology, but by the recognition that such scientific truths are only limited (there is much they can't explain) and provisional (they are always being amended) and that many new technologies degrade, even threaten, our environment and our lives. At the root of this postmodern change is technoscience. Its great fruit is cyborg society." Chris Hables Gray sees "Our Future as Postmodern Cyborgs." He notes "Many humans are now literally cyborgs. A 'cyborg' society has developed where the intimate interconnections and codependencies between organic and machinic systems are so pervasive that whether or not an individual in that society is a cyborg they live a cyborgian existence."

The critical issue facing humanity, according to Donald B. Louria, M.D. is "The Spectre of Emerging and Re-Emerging Infection Epidemics" which are certain to come in future decades.

"The critical issue is how to minimize their frequency and severity. The current focus is on global surveillance and immunization. It is the author's contention that the most effective approach is to ameliorate the societal determinants that create the milieu in which these infections arise and propagate. The two overarching societal determinants are population growth and global warming. Mitigation of societal determinants will require a sense of urgency accompanied by individual and political actions as well as educational change.

Clifton E. Anderson in "Genetic Engineering: Our New Genesis" says that "With positive guidance, molecular genetics may bring the world to a resource-conserving, life-affirming stance—a new genesis."

"Gene manipulation and transfer of genes from one organism to another gives scientists world-changing power. Plants, animals, and microbes can be altered and new life forms created. Working with people, geneticists have gained control over some inherited diseases and are studying far-reaching, hazardous methods of human gene therapy. The 21st century will feel impacts—positive and negative—of genetic engineering."

For Professor Allen Tough "Making contact with another civilization in our galaxy is likely to be the most important positive event of the coming millennium. The scientific search for intelligent life in the universe is rapidly expanding, and contact with at least one form of intelligence is likely within the next few decades." In "Making Contact..." he suggests "Four sorts of long-term consequences are likely to result from this contact: practical information; new insights about certain 'big questions'; a transformation in our view of ourselves and our place in the universe; and possibly participation in a joint galactic project. Of all the positive events that occur during the next thousand years, interaction with another civilization is to have the most profound and pervasive impact on human civilization."

The new millennium may witness a theological transformation in the conception of God, according to Robert B. Mellert in "The Future of God." "The death of God has been proclaimed several times in the last two centuries, but surveys tell us that in America today belief in God is stronger than ever. What is not so obvious is the transformation that has occurred in the conceptualization of God."

In the course of his essay he "reviews briefly some differing images of God in the Western tradition, as well as some of the philosophical and theological problems inherent in these images. A look at contemporary cultural assumptions suggests that our idea of God may still be in transformation: from an omnipotent, transcendent being into a concept that represents a more all-inclusive, pantheistic image. Curiously, this image may actually have some affinity with such disparate contemporary beliefs as fundamentalism, 'new age' thinking, and atheism."

"Do Universal Human Rights Imply the Future Development of a World Religion?" is the question posed by Charlotte Waterlow. She outlines "the United Nations' Universal Declaration of Human Rights of 1948 and the United Nations' two Covenants and the Council of Europe's two conventions which stem from the Declaration. For the first time in history they establish a legal ethical system for the world based, essentially, on the right of every person to the conditions of life to enable him or her to develop their potential."

She then discusses "the ethical basis of these Rights and outlines 'The Global Ethic' formulated by the Parliament of Religions which met in Chicago in 1993." She "suggests the traditional civilizations were based on the traditional religions, which laid down fixed rules of ethical behavior for government and social and economic life; as a result, these civilizations ticked over, essentially unchanged, for millennia. The Modern Age, based on "Liberty, Equality, Fraternity"and science, burst forth in Western Europe some 200 years ago, and in the 20th century is spreading across the world. It has opened the door to the development of ethics and religion based on the development of the individual person, to the emergence of a new kind of global society based on change, progress, and experiment. But in so doing it has unleased daemonic forces expressed in a backwash of emotional and intellectual fundamentalism. "...Out of this chaos is emerging the world religion of the future," she believes: "cosmic consciousness, based on Love as a spiritual experience, and expressed 'community' or fraternity."

Growing aging populations and the future are the focus of Dr. David Macarov's "Grow Old Along with Me. The Best (and) Worst are yet to Be."

"Lengthening life expectancy will almost certainly result in a continually growing number of older people. The result will be both positive and negative for the individuals concerned and for society as a whole. Most of the future elderly will be healthy, but those afflicted with degenerative diseases, care will be necessary for decades longer than presently. Those with enough income to cover their needs will be happy in their retirement, but the outlook for the poor elderly is bleak. Finally, there will be changes in family structure as three, four, and even five generations co-exist."

In contrast to the other authors, William Crossman raises a totally different subject—a very controversial one. His thesis; "CompSpeak 2050: How Talking Computers will Increase Oral Culture by the Mid-21st Century."

"In the 21st century, VIVOs (voice-in/voice-out computers using visual displays but no text) will make written languages obsolete. Written language is essentially a technology created 6,000 to 10,000 years ago for storing and retrieving information. VIVOs will perform this same function more easily, efficiently, and universally without requiring people to learn to read and write. There will be no compelling reason for schools to teach literacy skills. By 2050, a worldwide oral culture will be in place. Today's push to develop VIVOs is a further step in the human evolutionary drive to move past written language's limits and return to the biogenetic, oral-aural, pre-alphabetic roots of human communication and information storage. Young people's choosing TV, telephone, stereo, and computer games over books, letter-writing, etc.—and the resulting literacy crisis that engulfs our schools—is not the result of mental laziness or poor schools, but is an irreversible symptom of this deeper evolutionary process. VIVOs will transform every area of human activity in the 21st century, including education, the arts, human relations, politics, and business. Billions of nonliterate people, using VIVOs, will finally be able to access the world's stored information—if they can gain access to VIVOs. Access to VIVO technology looms as a key human rights issue of the 21st century."

Ambassador John W. McDonald calls attention to an additional aspect of the global chasm between the Haves and the Have-Nots: "The Global Impact of Information Technology: The Connected Versus the Unconnected."

"There are some 100 million people in the world linked to the Internet. Ninety-two percent are in the developed world. The remaining 8 million users are spread throughout the other 5 billion people in the world. This situation must change in the decades ahead. He discusses how the United Nations, led by Third World leaders, was able to create new institutions in the 1960s to force the rich North to focus on the economic development problems of the poor South. The world is faced by a similar crisis today; the divisions between those connected to the Internet versus the unconnected. The developed world must begin to rise to the challenge today. Several next steps are discussed [in his paper]."

Global "food security" is another subject of concern as we contemplate the new millennium. "At the close of the 20th century, although enough food is available to meet the basic needs of each and every person in the world, about 820 million people are food insecure and 185 million children are malnourished. If the global community continues with business as usual, prospects for global food security remain bleak with as many as 150 million children—one out of four children—still malnourished in 2020. Moreover, a number of factors emerging that suggest that humanity is entering an era of volatility in its food situation," report the joint authors of "Prospects for Global Food Security in the 21st Century," Per Pinstrup-Andersen and Rajul Padya-Lorch.

"With foresight and decisive action, we can create the conditions that assure food security for all people in the coming years. The action required is not new or unknown, but it calls for all relevant parties—individuals, households, farmers, local communities, the private sector, civil society, national governments, and the international community—to work together in new or strengthened partnerships; it requires a change in behavior, priorities, and policies; and it requires strengthened cooperation between developing and industrialized countries and between developing countries."

"Information as society's nervous system is a strategic component of every technological function. It is now America's largest economic sector. Benefits are compelling but, as with all technologies, there are unwanted consequences. Electronic communications at high speed and high volume cater a staccato of images that alter human attitudes and behavior and change how we think and what we think about. Attention spans wither. Dumbing down is accompanied by numbing down." This is a special area of concern to Edward Wenk, Jr. in his essay "Socio-Psychological Aspects of Information in a Democracy."

"Television caters news as entertainment or gossip, not enlightenment, focusing on the urgent and not the important. Programming succumbs to commercial pressures. The medium alters both the message and the messenger and befuddles critical thinking. With the Internet, authenticity of content and source is uncertain. Transmissions devoid of eye contact and body language lead to isolation and anxiety. To these add vulnerabilities in electric grids, banking and air reservations from sabotage or illicit use, or simply system crashes and the Y2K problem.

"Threats to democracy may be the most serious. Information technology is about power, economic and political. Corruption is fed by campaign financing from the imperative of candidates to buy TV time. Meanwhile, America's shared vision is fractured, and more communications have not heightened a sense of community or the quality of civil discourse. Impeachment proceedings confirm how intertwined are technology and democracy. Concerns are underscored that in the American future, democracy may suffer."

Two of our authors, Professor W. Warren Wagar and Dr. Robert Muller, Chancellor of the United Nations Peace University and former UN Assistant Secretary General, both believe that an efficient, proper world government is not a utopian dream about the future but a practical necessity.

Wagar's essay "Utopias, Futures, and H.G. Wells's Open Conspiracy' focuses on *The Open Conspiracy* which was "first published in 1928 and remains today a landmark in the history of utopias and of normative futures inquiry. This essay offers a new reading of Wells's book, viewed as a narrative of how humankind can make the perilous transition from the international anarchy of the 20th century to a system of efficient world governance. It concludes with some thoughts about the similarities between the time when Wells was writing and our own time, noting that his message is still a relevant and powerful prescription for world integration."

For Dr. Muller the new millennium may be the occasion for the realization of one of modern humanity's "impossible dreams"—some form of world government. In "The Absolute, Urgent Need for Proper Earth Government" he outlines a number of steps to achieve this goal as a result of "50 years of service in the United Nations system..."Some of the steps: "hold a World Conference on Proper Earth Government through the Free Market System...", "...a New Philadelphia World Convention for the Creation of a United States of the World" and "*a World Conference on Proper Earth Government through what the World's Religions have in Common...*"

In "The Next 1000 Years: the 'Big Five' Engines of Economic Growth" according to Graham T.T. Molitor, "Over the impending 1000 years, humanity will accommodate at least another five waves of economic dominance: leisure (dominant sphere of activity by 2015); life sciences (By 2100); mega-materials (2200-2300); new atomic age (2100-2500); and new space age (2500-3000). Each of these entrepreneurial sectors has been building momentum for hundreds, even thousands of years. Sudden surprises should not catch forecasters unaware of impending 'economic centers of gravity'."

THE CREATIVE UTILIZATION OF HUMAN CAPITAL

In John Nance's interview with Charles G. DeRidder's "Creating Tomorrow's Dream Team", we have a discussion of the importance and value of the Ned Herrmann Brain Dominance Model in the effective use of human capital.

"As we enter the next millennium, DeRidder says he is convinced that this process [the Herrmann Brain Dominance Index] provides strong evidence that we can better utilize our intellectual capital. It will not only improve team performance on the job and thus increase productivity, but it will also enhance the quality of life of the individual. This is, if a person who has a low thinking preference for a particular task is placed on a job that requires high thinking preference for the task, the individual's performance will be relatively low. Conversely, if the requirement of the job and an individual's thinking preferences align themselves, you have the potential of a world class performer who will most often give outstanding performance.

"DeRidder sees people working at their individual best who will find deep satisfaction in doing something they truly enjoy! When you see such people at work, he says, 'it's obvious that they are happier, more relaxed, and more successful in the ways that count.' Leaders and managers who are adept in using the brain preference process will be able to help their employees identify work opportunities that better match their personal brain dominance. Just possibly employees will cease to view their work as simply as 'job'. The potential now exists for the individual employee to pursue their life's work through their work life.

"DeRidder says, 'I'm convinced that the organization's intellectual capital base will be increased in value because of their enhanced wisdom, their judgement, and their inventiveness. I've seen it happen time and again."

"If we are to enjoy the full benefits of the information technology that is increasingly changing our world, societal and institutional changes will have to be effected." In his essay, "The Need for New Paradigms, " John Diebold offers the reasons for this need.

"A dual market system that can accommodate both large, global firms and small, entrepreneurial companies will develop. Some of these changes are already taking place, but many more alterations in the way we work and live, the way we think and learn, and the way our public and private institutions operate will occur early in the 21st century both as a result of the use of information technology and in order to benefit from the availability of this technology."

In "Three Parallel Revolutions: Minding the Economic Laws of Knowledge," Professor William E. Halal "draws on his Forecast of Emerging Technologies, examples from progressive organizations, and a survey of 426 executives to show that the laws of knowledge are driving three parallel revolutions. Information technology is accelerating breakthroughs in all scientific fields, producing a broad Technology Revolution that promises to transform the entire society over the next two decades. Exploding complexity wrought by this onslaught of change is creating an Organizational Revolution as business and government devolve into 'internal markets' of autonomous entrepreneurs who collaborate with clients and business partners to form pockets of 'corporate community.' With power and action now flowing from the bottom up and the outside in, a Leadership Revolution is also underway as CEOs and politicians are forced to give up their authority and govern primarily using vision, wisdom, and spirit. These three parallel revolutions collectively form a new global order that extends our traditional ideals of enterprise and democracy down to the grassroots."

FUTURIST OBSERVATIONS ON A NEW MILLENNIUM

In "After the Party is Over: Futures Studies and the New Millennium" Professor Graham H. May observes "The millennium offers Futurists an opportunity to place their concern with the future at centre stage, but it also has the potential to create a number of problems for us. The collection of year 2000 studies can be a useful learning vehicle and a stick to beat Futurists with. The buildup of interest in the media and elsewhere may be deflated after the millennium. The many popular forecasts that will be made, many of the which will not come true, may be used to question the validity of Futures world. Western Futurists should also remember that the millennium is founded in Western, Christian culture and may not be relevant to most of the world's population. Some suggestions are made to reduce the difficulties and maximize the benefit to Futurists of the millennium.

"Institutions of Foresight (IOFs) are purpose-built organizations that focus on one or another aspect of futures work. Depending on definitions there are, perhaps, several hundred around the world. Some of these are fully viable, while others no longer exist." Professor Richard A. Slaughter's "Creating and Sustaining Second-Generation Institutions of Foresight" suggests that both successes and failures provide useful pointers for creating and sustaining second-generation IOFs. In particular, the paper considers some of the implications of the Australian Commission on the Future (CFF). It looks back at the 12 years of its existence, attempts to summarize its achievements, and then suggests some lessons, or broad design principles, that may be useful to other such initiatives around the world.

Within the immense range of possibilities for the human race both for good or evil, one idea may become increasingly appealing and hopeful. With a new century—a prelude to new millennium—it may become desirable and possible to experience an "enlightenment" for the entire human family, and lay aside the burdens of past historical wrongs among all peoples and achieve the maturity to forgive past mutual cruelties, injustices, and antagonisms and manifest the wisdom to forget them.

The past is gone. We are powerless to alter it. Why should we remain slaves to its tyranny? The future is another matter. It is an open field of possibility. With forgiveness and forgetting, humanity can usher in the dawn of hope for all of earth's peoples. Such a change of mind, hearth and will could give rise to a universal ethical code honored and respected by all. Such a development could make "Earthlings" truly ready for the contact with other worlds. While there is such hope there is life.

LIKELY DEVELOPMENTS WITH UNIVERSAL RAMIFACTIONS IN THE NEW MILLENNIUM

FROM NOOSPHERE TO THIRD MILLENNIUM TECHNOSPHERE

by

Joseph N. Pelton

INTRODUCTION

Teilhard de Chardin is the person who probably had the first important insight about what the Renaissance implied for the future evolution of the human species. This French philosopher noted that world intellectual development was changing global society in profound ways and at increasing speed. He sensed that the "passage of time and the accumulation of knowledge" had begun accelerating and that something, which in today's lexicon might be called a "Global Brain" or a "Technosphere", was beginning to emerge. In Chardin's world of the late Renaissance the spread of information and shared human wisdom, known as the Enlightenment, seemed a most wondrous thing. He gave a term to this phenomenon--The Noosphere.

Just as the scientific community of his day comprehended that humanity lived within an umbrella of life-giving atmosphere, Chardin suggested that the spreading body of scientific and intellectual information on a planetary scale should be called the "Noosphere"—the expanding sphere of human knowledge and wisdom. He marveled at the birth of this Noosphere which seemed to have spontaneously arisen from the Renaissance. Perhaps even more amazing, this Noosphere seemed to be growing and expanding exponentially in volume and speed.

We know today that this same explosion of information as fueled by advanced communications satellites, fiber-optic cables, super-computers, artificial intelligence, expert systems, high density memory systems, bio-technology and advanced chip technology, is far from slowing down. Indeed we have moved well beyond Chardin's "Noosphere" to what might be called at the dawn of the Third Millennium the "Technosphere." How does one explain the forces driving the creation of this new "Technosphere"? Even more importantly how does one sort out and explain two extremely puzzling issues which seem to be at odds with one another?

Joseph N. Pelton *is a professor at the Institute for Applied Space Research at George Washington University, and director of the Arthur C.Clarke Institute for Telecommunications and Information, Washington, DC.*

Future Compression and Super Speed Evolution Which Now Seem Driven by a Mandate to Achieve Maximum Economic Throughput:

What are the social, economic, political and cultural implications and problems generated by this phenomena of Future Compression? In particular, what remedies may be needed for humanity to survive in the "Third Millennium Techosphere" if rates of information and economic evolution continue to increase their rate of acceleration? Does continued exponential expansion of the Noosphere carry potential threats to the survival of the human species?

Chaos Theory and Its Support for Human Evolution According to The Fourth Square-Root Law and a New Efficient Pattern of Sustainable Development

It is important to explore what new research in chaos theory, fractals, "artificial life" evolution and information theory tells us about patterns of evolution and development for all intelligent life. Such understanding might help us understand the future course that humanity will follow and what patterns the Technosphere might develop and exploit in the new Millennium.

When Did the Technosphere Begin and Where Is It Driving Us?

Humanity has been around on Planet Earth for somewhere between three to five million years. For a very long time—millions of years—very little seemed to have happened. If one views the history of human kind to be only a single day long, then the birth of farming and living in towns and cities occurs at 11:57 pm—three minutes until midnight. The industrial age comes at five seconds before the end of the day, while the age of computers, lasers, satellites, INTERNET, liposuction and spandex represents only some 700 milliseconds.

If we take the history of humanity and let it take the form of constructing a building which is 10,000 stories high—in the best traditions of the Tower of Babel—we can visualize this structure in two dramatically different ways. If we create the Building of Time, which is based on historial chronology, we find that the age of hunting and gathering represents 9.998 floors. Incredibly, the age of farming and permanent settlements represents only the top two floors. The Renassiance is less than 10 inches, very near the ceiling of the last floor, and the age of cyberspace is less than a half inch at the very summit of the Building of Time. (See Figures 1 and 2)

Now let's start all over and construct the Building of Human Knowledge and Wisdom. Suddenly the image is crazily reversed. The millions of years that represent the time of hunter/gatherers is only one story in the Building of Knowledge. The Renassiance is at most the fifth story. More that 9,000 stories represents the knowledge acquired and stored since World War II. Since the time of ancient Greece, as human population has increased by more than 55 times, the amount of information available on our planet has increased more than 10 million times--some suggest that the increase is almost 100 million times. By even conservative estimates the likelihood is that information is mushrooming at a rate that is some 300,000 times faster than our population growth.

FIGURE 1

BUILDING OF HISTORY

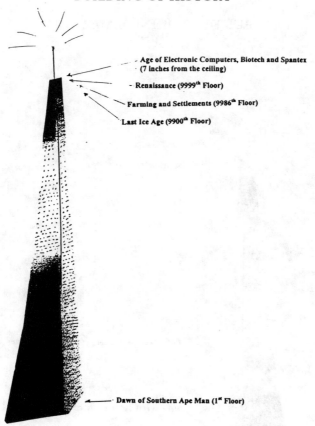

- Age of Electronic Computers, Biotech and Spantex (7 inches from the ceiling)

- Renaissance (9999th Floor)

- Farming and Settlements (9986th Floor)

Last Ice Age (9900th Floor)

Dawn of Southern Ape Man (1st Floor)

Nobody has yet explained where the "Noosphere" or our contemporary "Technosphere" truly started, but by the late Renaissance it had already reached a measurable speed, and now it is streaking forward at a blazing pace. It is no longer a second level exponential. Rather the rate of acceleration is increasing. What is called in physics quite simply "Jerk!!!"

We suspect that technology and the ability to grow crops in a permanent settlement which afforded people the opportunity to create specialized skills was a key ingredient. Likewise, written language, books, manuscripts and "other intellectual prosthetics" that followed (including today's computers and software) helped to create the modern Technosphere in which we now live. The parts played by culture, economic enterprise, religion and genetic diversity somehow complete the picture of how this history of information and communications and learning unfolded with quickening speed in the last few decades.

FIGURE 2

BUILDING OF INFORMATION

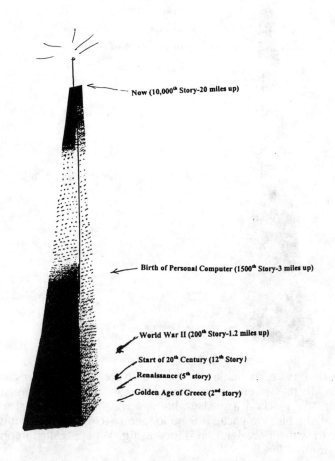

Now (10,000th Story-20 miles up)

Birth of Personal Computer (1500th Story-3 miles up)

World War II (200th Story-1.2 miles up)

Start of 20th Century (12th Story)

Renaissance (5th story)

Golden Age of Greece (2nd story)

Many scientists, technologists, engineers and historical determinists such as Buckminster Fuller would suggest that this process was not only inevitable but that the resulting technocratic society is desirable, just and ultimately "good". There are others such as Jacques Ellul, Victor Ferkiss, Lewis Mumford et. al. who disagree with such a "rosy" assessment. They point to the problems of urban decay, high rates of criminal activity and drug addition, proliferation of nuclear weapons and biological and chemical weapons, techno-terrorism, the widespread lack of education and health care services for some 2 billion people, problems of religious intolerance, racial bigotry, and even genocide. There is a good deal to suggest that each technological development such as automation leads to positive and negative impacts such as efficiency and reliability on one hand versus unemployment and pollution on the other.

A review of life in the Technosphere shows key problems such as: environmental pollution, decertification, global warming, proliferation of weapons of mass destruction, information overload, stress, drug abuse and addiction, loss of privacy, and "deskilling" of many occupations. Jobs which once required a high level of training or education (including architecture, accounting, musical arrangements, property appraising, etc.) are under threat to artificially intelligent programs and expert systems. There are other problems such as how much will advanced information systems really help to improve basic education, health and medical care, governmental services, etc. Communications and information technologies were supposed to provide key breakthroughs but, in fact, when objective measures are applied, only marginal gains have been achieved. In fact, to use education as but one example, in most of the industrial countries average student test scores are declining rather than improving despite sophisticated instructional technology.

Technological progress and strategic research programs, which are the products of today's Technosphere are producing many undesirable social, cultural, and economic results for several reasons:

A. Outmoded Economic Throughput Concepts Born of an Industrial and Manufacturing-Based World Which Are Inappropriate for the Age of the Technosphere

The industrial-based system of maximized economic throughput is based disproportionately on capitalist goals of wealth generation rather than on survival and evolution of the human species. The main fallacy here is an outdated socio-economic system which maximizes more and more specialization and reduction of people's role in economic systems to "cogs in a machine" without overall meaning or value. This same economic goal system also promotes increased separation of the "production" side of the economy from the "consumption" side. This increasing separation of production and consumption is based on the theory that this will produce maximum net positive economic gain and efficiency. This is not necessarily so. In biological systems tested over eons, redundancy ultimately provides "survival margin" for the species. But in late 20th century economic systems "redundancy" and even "esthetics" are equated to waste and mismanagement. Cyberspace economics of the 21st century will eventually prove to be much different from the "industrial economics" of the post World War II era.

b. Fundamental Realignment of Social Evolution in a Technosphere-Based World

Perhaps it is only an accident that a fundamental shift in the course of human evolution is occurring at the turn of a millennium, but the scope and the range of the coming changes should not be underestimated. If the human species is to achieve success as a species, i.e. continuous development for at least an eon (i.e. one billion years), the most severe test will come within the next century. Fundamental problems with global warming, pollution, preservation of the ice caps' albedo, global health and education, nuclear disarmament and even world-wide zero-population growth will need to be solved in the next century if the species is to have any chance at longer term survival. The pace and character of our economic systems must change to support new value systems. Even more fundamentally the way we build our Technosphere will need to adapt to "higher-level forces and value systems" as well as powerful "biological laws of evolution" that are only now beginning to be unlocked in the research results of the Santa Fe Institute. This innovative group and its exploration into fractals, chaos theory and non-linear math are giving us totally new views of our world and how it works.

Nobel Laureates at the Santa Fe Institute and other interdisciplinary scholars around the world are trying to see how economics, physics, biology, mathematics and information theory relate to each other in the world of "non-linear" mathematics and chaos theory. Their results show quite startlingly that the evolution from non-life to living cells, to multiple cell organisms, to complex life forms, to mammals, to humans, to human settlements and villages and to planetary cultures seem to have some innate formulas and rules built into the evolutionary process. These "living patterns of evolution and development" may, in fact, be so fundamental as to explain what is meant by such emerging concepts as trying to "overcome entropy". Entropy, of course, is the fundamental nature of all systems, inert and alive, to become disorganized and to breakdown toward chaos. For better or worse, a fundamental change in world social patterns may be about to unfold if our new theories of life prove correct.

USING THE TECHNOSPHERE TO SOLVE GLOBAL PROBLEMS

A. The Transition from 20th to 21st Century Value Systems

The Twentieth Century, despite pockets of totalitarianism, has marked the ascendancy of democratic political systems that were based on a free-market, capitalistic-based society where maximization of profits and access to material goods and services have almost seemed to be an end in themselves. In the 21st century as human economic wealth extends across the planet and health care and education become more universally available, several key results can be projected.

- **Stable Global Population**

First, United Nations studies project the gradual slowing of birth rates that will result in virtual zero population growth on a planetary level between 2050 and 2075.

- **Sweeping Implications of Tele-Commuting in an Increasingly Global Service Economy**

Second, the increased spread of tele-commuting in service industries will serve to connect not only rural and urban areas, but also promote "electronic immigrantion" so that qualified teleworkers can roam the planet. As the number of tele-commuters moves above 100 million in about 2015, this will serve to distribute world-wide wealth more rapidly, start to impact global pollution, change patterns of real estate valuation, etc.

- **Declining Importance of the Traditional Nation State**

Third, the continued spread of global corporations, the increasing world-wide access to Cyberspace-based networks, plus the increasing predominance of prosperous democratic countries will all serve to diminish the force and power of the "nation-state", particularly among the world's most affluent and non-traditional societies.

• **Possible Rise of a New Techno-Value System**

Fourth and most significantly there will be the "opportunity" (possibly even the likelihood) of a new "techno-value" system. This possible new global paradigm would most likely spring from the new Millennium generations' concern about global pollution and the destruction of the planetary ecology. This would be a "win-win" value system that might well seek ways to advance global human evolution while still preserving the Earth's "biosphere". It would stress environmental conservation and sustainable development over individual material wealth, yet still see new ways to expand eonomic prosperity along new industrial lines. Under such a new value system we could still see remarkable new growth and advancement of new industries such as" recyclable energy" systems, "smart and clean" construction industries, "intelligent transportation" networks, new tele-education and tele-health systems, etc. Smarter and cleaner does not have to mean poorer or greater austerity.

B. Artificial Life: Interpreting the Meaning of Fractals, Chaos Theory and Non-Linear Math

It may well turn out that the most important development of the 20th century will come from, of all things, applied math. There is ever growing evidence that "God's Math" thinks in terms of fractals, chaos theory and non-linear math. The patterns of coastlines, stock market behavior, weather, and even the evolution of life seem to flow from our latest understandings of non-linear math and chaos theory. The latest studies from the Santa Fe Institute, where interdisciplinary applications of non-linear math and fractals are over a decade old, reveal dramatic new patterns among all of the animal and plant kingdoms. Recently it has been shown that the circulatory systems of animals and plants have much in common and that evolutionary patterns build toward larger, more complex and "efficient" systems.

The newly revealed "Fourth Square-Root" concept shows why metabolism rates and blood circulation become more and more efficient in larger and larger organisms. It turns out for instance that mice, men and elephants all have metabolism rates that conform to a universal "fourth square-root" law that translates to the plant world as well. Most complex animals from mice to elephants for instance live for about 1 billion heart beats despite their radically different lifetimes and because of "efficiency scaled" metabolism rates and circulatory systems. (See Figure 3)

FIGURE 3

EFFICIENCY SCALING IN LARGER AND MORE COMPLEX ENTITIES

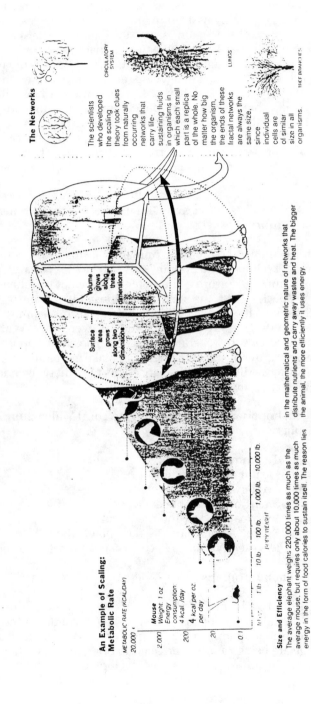

Other studies from the Santa Fe Institute have focused on a complex computer model of evolutionary behavior from complex amino acids, to amoebas, to multiple-cell creatures, to reptiles, to mammals to humans to human civilization, to the stucture of the Universe. These again seem to follow fractal-like patterns of behavior. Increasingly sophisticated insights about how life evolves almost seem to suggest that we humans may not really have as much choice about how we develop as we previously thought. The combined studies of Charles Darwin and the new chaos theorists point out that evolutionary development may not end with speciation, but rather may extend into "societal patterns of progress" as well. If so, there may well be some hope that humans can evolve back from "excessive patterns" of development that value growth over survival, and thus create a new ethos of "sustainable development."

C. A New Thesis: Six Key Dimensions of the Technosphere in the 21st Century

If it is indeed possible for humanity in the 21st Century to evolve a new "Techno-Value System" this will be the critical step to survival as a planetary culture. The potential of what can then be achieved as a species over the next one thousand, the next million and ultimately the next billion years seems almost unimaginable—perhaps even entropy could be overcome. Within this new planetary culture several key technologies and applications within the ever-growing Technosphere can make extroadinary contributions to this new Millennium Society. This new thesis is indeed revealed in the acronym T.H.E.S.I.S. as follows:

An Important THESIS Within the Third Millennium Technosphere

- Telecommunications
- Health and education
- Ecology and environment
- Smart or recyclable energy
- Information and Internet Next
- Space Applications (i.e. Navigation, GIS, Imaging, Disaster Warning/Recovery and Planetary Safeguard)

Telecommunications

Currently advanced telecommunications systems are operating at fast digital speeds of up to 10 gigabits/second. In the laboratories systems that operate in the 100 gigabits/second to the terabit/second range are being rapidly perfected to bring into operational use in the new millennium. The latest developments in fiber optical cable, radio frequency-based mobile and satellite transmission systems and sophisticated multi-plexing technologies suggest that in the 21st century almost unlimited amounts of information will be able to be relayed almost anywhere at any time and at incredibly low costs. Local, regional, national and global telecommunications will essentially cost the same even though tariffs will be based on the greater value of long distance communications.

The implications for the Technosphere are many fold and potentially profound. Increasingly we will move ideas and information much more so than people and physical products. The implications for a 21st century Technosphere will be almost everywhere. In a world in which most jobs are in the form of services we can and will tele-commute to work across town or even across the world. We may see as many as 50 million tele-workers by 2010 and over 100 million by 2015. By the end of the 21st century the majority of all jobs may involve tele-workers. Not only will we have "electronic immigrants" who travel across the world to work, but also "electronic tourists" and even "electronic soldiers" who engage in tele-wars.

Although there will still be those who are denied access to all the information they want and need, there will be the reverse problem of information overload and the need to filter the gushing Niagara of "passive information" that will spew forth from a vast array of information machinery. By 2050, the global population will be between 8 and 9 billion. At that time each and every man, woman and child on the planet will each be able to access an array of information equivalent to the Library of Congress (i.e. 10 terabits of information). This in itself is not remarkable. It is the fact that everyone on the planet could do so without necessarily overlapping the information of anyone else that is at once impressive and frightening. In other words there will be by mid 21st century a global information system that is somewhere around 80 to 100 sextillion bits of information.

Our vast information networks of fiber, wireless and satellite systems will be able to transfer data, images and conversations around the planet at will. These systems will be able to move the equivalent of a large book at less than one cent in cost.

Even so, the issue of disparity of access to information will remain. Thus while some will pay more to filter out the "active" information they desire from "passive" information they do not want, there will still remain developing societies with inadequate and unreliable information networks. Bridging the global information and telecommunications gap will be a key challenge to devising a successful Technosphere that meets global needs in the next century.

Health and Education

The huge size and throughput capabilities of 21st century telecommunications and information networks created to support economic and commercial activities will be awesome—100 to 1000 times the size and speeds of those we know today. Given the existence of such large and low cost information networks, it will be increasingly viable and cost-effective to deploy networks to support health, medical and educational services. The idea of an electronic tutor that has encyclopedic knowledge of dozens of subject matters and which can download instructional information and provide interactive assignments in dozens of languages on demand at a cost of one penny per course is certainly realizable in less than two decades.

In China today there are more than 4 million students receiving instruction via the National Satellite Television University. In India there are more than 1 million students who are receiving course instruction via the INSAT satellite network. Entities such as the National Technological University, the Knowledge Network of the West (KNOW), the TeleKollegg of German, the Open University of the United Kingdom, the University of the South Pacific, the University of the West Indies, and many others have proved that tele-education services linked to and reinforcing traditional educational systems are sound ways of improving the quality, speed. currentness of information and effectiveness of learning systems.

The evidence in tele-health care systems is less extensive and certainly less well documented. Nevertheless, tele-health projects such as those carried out around the world over the last two decades have shown impressive results. The Memorial Hospital of Nova Scotia for instance has shown how it can electronically serve not only the rest of Canada but many other parts of the world as well. This and other hospitals in dozens of countries are showing how the range and depth of tele-health care services can continue to expand and handle an increasing breadth of health and medical problems. In the 1970s and 1980s health education such as nutrition and basic education to address infant care were the main thrust of satellite-based remote medical services. Over time tele-health services have expanded in locations such as Canada, Alaska, Australia, the Caribbean, East Africa, etc. to include medical diagnosis and testing, and even support for minor operations. At the most advanced and exotic stage there are projects such as NASA's bio-medical research team created to explore such techniques as remote tele-robotic surgery. In this case a laser beam guided by a robot computer is able to track the commands of a remotely located surgeon down to one-thousandth of an inch tolerance. Today the types of remote tele-health services range from very basic offerings costing only modest amounts of money and involving a minimum of electronic equipment up to multi-million dollar activities at the highest level of sophistication.

This much is clear. There are today some two billion people out of a global population of 5.7 billion who have either no or very limited access to schools, health care clinics or doctors. The only way that a significant change can be made in providing health and education services to this vast underserved population is to utilize the best of the tele-education and tele-health based services in concert with traditional programs. This wide-spread provisioning of tele-education and tele-health programs is most likely to occur as an off shoot of the advanced information and telecommunications networks that will be deployed in the 21st century. Ultimately, systematic and high quality education and health care services for the entire planet will involve more than access to electronic equipment and networks--this is, in fact, the easy part. The challenge will be to develop the content, create in the necessary languages the instructional materials, train the teaching aides, provide the paramedics, and generally create the human infrastructure and training materials to make these tele-services truly effective. Even so, the objective of complete planetary literacy and uniform health care for a stable human population of perhaps 10 to 12 billion people would seem a viable objective for a 22nd Century Technosphere.

Ecology and Environment

The concept endemic to 19th and 20th century industrialization was frequently expressed as "Man's Ability to Conquer Nature." As we begin a new millennium there is increasing recognition born out of the efforts of such individuals as Jacques Cousteau, Lester Brown, and others that we humans must live in harmony with nature. This means that we must derive energy from the sun, wind, the oceans, geothermal wells or artificial photosynthesis. This means that we need to build our houses, our buildings, our automobiles and other consumable products out of materials that we can grow or recycle endlessly over time. It means that processes that pollute, stain, heat or cool our environment must be eliminated or marginalized. Oil must not freeze into our polar caps. Tropical rain forests, wetlands, and mountains must be preserved and human settlements contained to finite regions of the globe. Once we set the key parameters for sustainable global development and then work backwards, we can invent a prosperous future for humanity that allows our species to survive an eon rather than a few hundred more years. The essense of the Telesphere THESIS is thus in the goal of a sustained ecology or environment that can stand the test of time. The initial Club of Rome and International Institute for Applied Systems Analysis (IIASA) studies of the 1970s which tried to devise longer term strategies for human survival and sustainable development were admittedly naive and flawed, but they marked an important depature. With the new telecommunications, information systems and artificially intelligent software, we have at our fingertips the potential of a new beginning. With the new millennium there is need to develop new economic and environmental goals that can prove politically viable, economically prosperous and ecologically attractive solutions to our long term survival.

Smart Energy

No single development of the Technosphere in the next century is more crucial to our positive development and survival than clean and recyclable energy. The success of the 21st century Technosphere will in large part be based on deriving energy from recyclable sources, conserving the consumption of energy, and using so called "smart energy" for heating, cooling, lighting, transportation, and in a broad spectrum of other applications as well. The conservation of energy not only is highly economic, but it also reduces one additional source of global warming. Today all our sources of energy ultimately come from solar illumination of the planet, but in time it could come from solar power satellites in earth orbit and even solar-powered space colonies, though this too has implications as a cause of harmful global warming.

In the past, virtually all crucial decisions about our global development have been market-based. On the basis of dollar per kilowatt, solar and wind power have been considered to be "non-economic". In the 21st century, it may well be that economic goals will be set so that by dates certain various new smart energies will be not only developed but engineered so as to be cost efficient as well. As cheap and plentiful sources of oil, gas, coal, and uranium diminish, the economic incentives to develop smart energy and cost effective energy transmission systems will ineluctably increase.

Information and INTERNET Next

The exponential increase in information will revolutionize the world. Economically, the world's gross domestic product will increasingly depend on information and services and less on goods and products. The nature, form and practice of education and training will be revamped. In this process thinking skills, hands-on problem solving, interdisciplinary team work and global interactivity will be emphasized, while memorization and storage of facts, over-specialization and traditional learning processes will decrease in importance. Likewise, mechanisms for accrediting of education will be revised to be more results-oriented, and degrees will need to updated and recertified. The idea of a static education that applies to only a 15 year period out of lifetimes that exceed 100 years will seem extremely quaint in an age of dramatic information explosion.

In the future we will see artificially intelligent knowbots that will actively filter quintabits of information to derive the information that a particular user seeks to know, screening out "passive information" that interferes with obtaining knowledge or wisdom. The "fuzzy logic" systems that comes from chaos theory and complexity will be critical to finding successful strategies for dealing with "too much information." One of the key issues will be how to store and retrieve information quickly and efficiently wherever we are. This strongly suggests that computer chip storage systems implanted in our skulls could well be part of an overly information-rich society. In this one might contemplate whether contemplation and creative thinking is more important to the longer term human condition than instantaneous and exhaustive access to raw information.

Space Applications

The synoptic overview and large scale coverage of space-based systems will undoubtedly increase in importance in the 21st century. Space-based exploration and application are revolutionary in several ways. Perhaps most important is the ability of satellites and space systems to allow us to macro-engineer our planetary environment in a way that terrestrial technologies do not. To monitor the ozone hole, reforest our tropical vegetation, map our countries, create a global navigation and control system for ships, aircraft and land vehicles, and generally manage our planet, the importance of space-based sensors, communications satellites and intelligent networks will only increase. In only a few short years Global Information System (GIS), satellite navigation, remote imaging, meteorological networks and mass consumer satellite networks have revolutionized the scope and reach of space applications. Today this space-based activity at about $50 billion a year is seemingly minor in a world with a $50 trillion dollar economy. But in terms of evolving to a new Techno-Value system wherein we create a Technosphere devoted to creating a sustainable global economy, there are few components that will be of greater importance.

Final Observations and Predictions

The path to the future is far from clear. Our emerging Technosphere with over a million times the capacity and information of that contained in Teilhard de Chardin's Noosphere is an extremely powerful tool. If this tool is used to sustain traditional capitalist economic values of maximizing economic throughput of material goods and services under a philosophy of endless economic expansion and consumption, then the survival of the homo sapien species must lay in serious question. If on the other hand we find a way to let human society move toward a true "Fourth Power Square" Law of Evolutionary Efficiency, then there could well be hope. In this scenario, the Technosphere would allow us to derive the new technologies, applications and services we need to be prosperous but also achieve true sustainable development. Here are some of the key challenges for which we must find good answers:

Managing Mega-Data and the Coping with the "Global Brain"

The increasing rate of acceleration in the production, processing, storage and communications of data is no trivial matter. As mere 64 kilobits/second data processing units, humans are in danger of being swamped by passive information and rendered obsolete by machine-to-machine communications networks. We have for decades now thought in terms of the McLuhanesque vision of the "Global Village", but in truth we are rushing toward the Global Brain. Exactly what the role of humans in the age of the "Global Brain" is and how we adapt to this new information environment is of critical importance. Only when we know the answer to this basic question will we know how to alter and improve our approach to education, health care, labor practices, employment, and culture, etc.

Privacy and Morality within the Technosphere

Like no previous time in modern history, the populations of free and democratic states today feel virtually no sense of privacy. Systematic surveys of citizens in the United States over the last 15 years has shown that the percentage of people who feel that privacy is a problem or a significant problem has risen from 65% to nearly 90%. The rise of technology and complexity in society has also seemingly given rise to a host of social problems and moral ambivalency that include rising divorce, drug addiction, child and spousal abuse, road rage, general stress, and mental disorder. Nowhere does technology seem to offer promising remedies to these problems. This must be considered an area of great vulnerability to the future evolution of the Technosphere. Some (myself included) believe that the increasing separation between production and consumption in the Technosphere is central to these issues and their future solution.

Human-Machine Interfaces and the 168 Hour Work Week

The advanced telecommunications and information systems of the early 21st century which are being designed to operate at terabit/second speeds are driven not by the needs of human communications but by machine-to-machine or human-to-machine links. Thus we will soon move to an environment where the preponderance of communication is between machines. Furthermore automated systems for billing, aeronautical control, industrial production, and trading of stocks, bonds and commodities will run on a 24 hour-a-day, 168 hour-a-week basis. As this focus shifts from human labor to non-stop machine driven operations, the whole cultural environment of humanity will be ineluctably changed forever. The economic, social, cultural and even political implications of this shift to a world of the "Global Brain" and the non-stop economy need to be far better understood than they are today.

Overcoming Entropy

If one were to try to define the ultimate human challenge it might likely be to "overcome the inevitably of entropy." It seems that only human intelligence and perhaps that of other intelligent beings in the universe are able to create some order out of chaos which entropy seems to be constantly creating throughout the Universe. Yet if entropy is constant and invariably present throughout the Universe then how is it that stars, solar systems and galaxies seem to have well-ordered forms and structure that follow the "rules of fractals." If humans were indeed able to understand the true nature of entropy and learn how to overcome it, then almost any feat would seem possible. In such an environment of enlightenment we might "terraform" Mars into a livable planet, create a new star to replace the Sun or colonize the Galaxy.

The Future of the Species

It seems likely that if homo sapiens or its successor (homo electronicus?) is to survive as a species (which is to say survive a billion years or more) then the 21st century will be the key period in long-term human survival. In the first century of the new millennium we will need to solve the problems of the ozone hole, global warming, a darkening albedo within the polar caps, large scale impact from meteorites or comets or even the more mundane issue of economic and industrial adjustment to a stable global population. The question is in the human quest to adapt and survive. Will we evolve, perhaps via genetic engineering, to create a new species? Perhaps even more importantly will we be able to evolve a new value system with this new Technosphere that promotes methods of survival and sustainable development, allowing our species to last another eon or more. Just as one example, there is increasing evidence that meteorite and comet impact on the earth have more than once threatened the biosphere. As Arthur C. Clarke has said, "...the dinosaurs did not survive as a species because they did not have an effective space program." If ever a species needed a survival plan it is likely to be that of homo sapiens in the age of the 21st century Technosphere.

OUR FUTURE AS POSTMODERN CYBORGS

by

Chris Hables Gray

What's this? Postmodern Cyborg? The worst of academic jargon married to a science-fiction monster? Why not talk about something real? Why not something important? Well, cyborgs are real, from granny with her pacemaker to the technology dependent astronauts in space, and whether you like the label postmodern or not it is clear to most historians, philosophers, farmers and cab drivers that the times they are a'changing. Drastically.

We don't live in the stable modern world our grandparents did. Their belief in inevitable, comfortable, progress has been supplanted by our realization that scientific and technological innovation are relentless and quite ambiguous. Our ancestors' acceptance of the natural limitations of space-time and life and death have been replaced by the fear and hope we feel about space travel, apocalyptic war, immortality, global pandemics, virtual community, ecological collapse, scientific utopias and cyborgization. The modern assurance that we humans control our own destiny has been blasted away by horrific wars, ecodisasters and a proliferation of new scientific discoveries and technological innovations that range from the sublime to the patently evil, as a few hours of television viewing or internet surfing can easily demonstrate.

At the root of all of this change is that great creation of the modern era: technoscience. I use this term advisedly, knowing it will annoy a large number of readers who like to keep their science and technology separated, at least conceptually. But while science and technology are clearly different things sometimes, they are also often mixed together in ways that are impossible to untangle. Their symbiosis is much greater than their parts and it is profoundly changing human culture. Humans have always been innovators and makers but starting around 500 years ago society began to institutionalize scientific and technological discovery. Since then those institutions have grown stronger and more effective and the small stream of new scientific understandings and new technological inventions has turned into a flood that shows no sign of slackening. In fact it grows stronger all the time. A majority of all of the scientists and engineers who have ever lived are alive right now and they are busy! The hardest thing about understanding this explosion of new knowledge and new things is stepping back away from it mentally to ask what its key features are. How is it changing? What is it doing to us?

Chris Hables Gray *is a professor at the University of Great Falls, Great Falls, Montana.*

I will first try and show why it makes sense to think of our times as post-modern. To situate ourselves historically is hard, but necessary. Unlike some historians I believe that history is actually useful. Where we've been is the best predictor of where we are going, yet it is far from infallible. What history can reveal are the choices we have. And we do have choices. My main point about postmodernism is the same as the theme of that great cyborg movie, *Terminator II*—"The Future is not yet written." This is the good news about the human situation generally. We make the future, not out of thin air of course, but out of the past and the present. And there is also good news about our specific postmodern condition: it is definitely not permanent. Postmodernity is transitory, it is a crisis, and the choices we make will determine what will replace it.

Then, as if explaining postmodernity weren't hard enough, I will try and convince you that the idea of the cyborg (the cybernetic organism, the merging of natural and artificial elements into one system, even into one human) can help us understand what we are and what we might become.

Let us begin at the end of history so far, at today, our postmodern moment.

POSTMODERNITY

The term postmodern is like a red flag to many a traditional scholar. I have to admit I resisted it myself at first. It seemed trendy and quite empty of real meaning. However, as I pursued my graduate studies into the history of technology and war I began to realize that the well recognized five hundred year period of Modern War had come to an end. I was not alone. Many new labels have been coined over the last few decades by historians, military thinkers, and philosophers to describe what Modern War had become, including: permanent war, technology war, high-technology war, technological war, technowar, perfect war, imaginary war, computer war, war without end, Militarism USA, light war, cyberwar, high modern war, hyper-modern war, hyperreal war, information war, net war, neocortical warfare, Third Wave War, Sixth Generation War, Fourth Epoch War, and pure war.

None of these were sufficient. Some recognized the special role of information machines and ideas in contemporary war but not one captured the full scope of just how much war had changed while retaining the insight that there had not yet been a total break with Modern War. Modern War could be differentiated from Ancient War by its commitment to technological and scientific innovation, by its acceptance of the complete mobilization of a society for war, by its belief in the usefulness of war as fundamentally a political instrument and by the combination of these factors in a drive for total war. With the advent of superweapons, the most obvious being the atomic bombs used at the end of World War II, total war becomes impossible, yet the other central ideas of Modern War live on. Thus we have postmodern war, full of paradoxes, struggling to survive.

I won't go into the rest of this story for it is in my book *Postmodern War*. But there are two important things that come out of this analysis that are directly relevant here. First, postmodern war is dependent on a new level of integration between soldiers and their weapons: human-machine weapons systems or, in other words, cyborg soldiers. It is actually by maximizing computerization and perfecting the warrior-weapon interface that many military analysts expect to make war useful again. This is discussed more below. Second, the rise of Modern War corresponds with the rise of the modern state, of modern science and with the spread of Western colonial systems throughout the world. The end of Modern War with the atomic bombing of Hiroshima and Nagasaki also marks the beginning of the end of European colonialism and profound changes for the modern nation-state and modern science.

And these changes mirror those usually ascribed to postmodernity. There is a proliferation of different, even contradictory, factors (bricollage); a collapse of a universal belief in single explanatory systems and ideas (the end of grand narratives); and a recognition of the centrality of information and its subcategories (simulation, computerization). In politics we see new forms of organization and complex loyalties, all in the context of a world knitted together by new communication and travel technologies. Science, which in one respect is fracturing into thousands of fields, is also uniting around informatics. Cybernetic principles have become central to most disciplines theoretically even as computers have become indispensable in practice.

When I thought of all these things together I could not deny the term postmodernism any longer. The implications of accepting that we live in the postmodern period are direct. It means that it is crucial that we understand what modernism was. It means that this is a transitory period bringing with it severe social and economic changes. And it means that primary engine of modernity, technoscience, is more important than ever and that we must confront its most important product: a relationship between humans and our technologies that can best be termed cyborgian.

THE IMPORTANCE OF THE CYBORG IDEA

The term cyborg is new but the idea is not. The dream of creatures combining artificial and natural aspects can be found as far back as the classical Greek and Indian civilizations. For example, when the goddess Demeter accidentally ate part of the shoulder of Pelops as the result of a mortal's experiment in the discernment of the gods, she replaced the shoulder with an ivory one. The earliest prosthetics known date from the same time as this story.

Five hundred years ago in Europe, during the birth pangs of the modern era, stories of golems, homunculi, and talking heads proliferated as did a new generation of elaborate prosthetics with movable fingers. Almost two hundred years ago Mary Shelly made up a warning tale about an arrogant doctor constructing a creature from the body parts of the dead. The monster in her *Frankenstein* was probably the first fully realized cyborg—organic but constructed, of human parts but post-human.

A hundred years ago scientists were discovering invisible light such as X-rays, engineers were analyzing workers as if they were machine parts and in the military and medicine great strides were made in rationalizing human behavior and the human body.

At the beginning of the twentieth century humans conquered the air, the deep sea, and even distance on land with machines. By World War II the military was planning in terms of man-machine systems (like those that made blitzkrieg possible) and the first artificial kidney was operating in the occupied Netherlands. After the war humans explored the depths of the seas and also moved out into space, both locations where we can only exist as cyborgs.

It was for a 1960 NASA conference on modifying the human for living in space that cyborg was coined by Manfred Clynes. Clynes, a world class pianist with a knack for inventing computers, melded cybernetic and organism into "cyborg" to enliven the ideas of his paper, coauthored with Nathan Kline, the famous psychiatrist and expert on psychotropic drugs. Clynes and Kline suggested that humans could be modified with implants and drugs so that they could exist in space without spacesuits. It is not as crazy as it sounds, but even Clynes would admit today that you'd need genetic modifications as well to make such a transition possible. In fact Clynes, who continues to work on cyborg ideas with his theory of sentics, (the physiological basis of emotions) and with a number of startling computer-music programs, now feels that humans will pass through at least four different cyborg stages, with genetic modifications being the last.

The term cyborg did catch on, but not among scientists who preferred more specific labels such as biotelemetry, human augmentation, human-machine systems, human-machine interfaces, teleoperators, and to describe copying natural systems to create artificial ones, bionics. However, cyborg took off among science fiction writers who had already recognized the incredible integration of technology into natural systems that was starting to transform society.

As with postmodernism, I initially resisted the argument that the figure of the cyborg was crucial for understanding contemporary society.

Although Donna Haraway's essay "Cyborg Manifesto" had encouraged me to apply to the doctoral program where she teaches, I wasn't struck by the cyborg idea (very familiar to me as a science fiction reader and writer) as much as I was inspired by her refusal of either an adulation or a outright rejection of technoscience. She argued that we have to start...

> ...taking responsibility for the social relations of science and technology means refusing an anti-science metaphysics, a demonology of technology, and so means embracing the skillful task of reconstructing the boundaries of daily life, in partial connection with others, in communication with all of our parts.

It was only when the cyborg continually reappeared during my graduate research on the role of computers as weapons and metaphors in military policy that I began to realize that Donna was right about the cyborg, perhaps even more right then she realized herself.

So after I finished my dissertation on postmodern war which included a large section on cyborg soldiers, I started researching cyborgs in earnest, tracking down Manfred Clynes and interviewing him, tracking down Jack E. Steele, who coined the term bionics, and interviewing him. I got a fellowship at Oregon State which gave me the time to carefully read the history of cybernetics. An NEH summer grant sent me to Case Western where I studied the development of prosthetics and the transition of medicine from a biological to an informational metaphor, something Donna Haraway was also looking at, especially in terms of the immune system. Then a fellowship at NASA's history office allowed me to explore human-machine integration in space exploration. During this period I put together a collection of the most important historical and analytical articles about cyborgs with the help of two colleagues, which we labeled *The Cyborg Handbook*. And so, before I knew it, I was convinced that the idea of the cyborg explained everything.

Well it doesn't, of course. That bit of mania has subsided somewhat. But I do think one can argue that the cyborg idea is invaluable for understanding who we are and what we might become. As a feminist philosopher I don't think we can ignore the body as the very ground for our interactions with the rest of reality. Ideas are fine, they are crucial, but they don't occur, or work, in a vacuum. To understand ourselves we have to start with the body.

The twentieth century human body can be conceived of through any number of rich and insightful metaphors. In important ways it is a disciplined body, a textualized body, a gendered body, and a resisting body. But more and more it seems to me that one of the most fruitful metaphors is to conceptualize the human body as a rhetorical and material construction of the discourses and cultures of technoscience, the mass media, and the military; a creature that combines informatics, mechanics, and organics: a cyborg.

Many humans are now literally cyborgs. Their inorganic subsystems can range from complex prosthetic limbs to the programming of the immune system that we call vaccinations. In the industrial

24

and so-called post-industrial countries a "cyborg society" has developed where the intimate interconnections and codependencies between organic and machinic systems are so complex and pervasive that whether or not any particular individual in that society is a cyborg, we are all living a cyborgian existence.

But haven't people and societies always been cyborgian in some sense? In a word, no. Certainly we can look back into the human past and note how crucial the human-tool and human-machine relationships have been, but quantitatively and qualitatively the cyborg relationship is new. Today there is the integration of the human and the tool; there is the symbiosis of the human and the machine. Yes, the cyborgian relationship is a direct development of these earlier human-tool and human-machine pairings, but we have now entered a fundamentally new stage, perhaps even a culmination, of this history. *Cyborg* is as specific, as general, as powerful and as useless a term as *tool* or *machine*. And it is just as important. Cyborgs are proliferating throughout contemporary culture, and as they do they are redefining many of the most basic concepts of human existence.

THE PROLIFERATION OF CYBORGS

Cyborgs are being created in numerous places in our culture. Some of these cyborgs are imaginary, some are quite real. Others seem to be both. A good place to begin a cyborg tour today is where my own research started, the military.

Lewis Mumford argued that the very first machine was an army, with the soldiers and their weapons making up the moving parts. This early proto-machine looks suspiciously like a cyborg, as does the 20th Century "megamachine" that Mumford railed against. Considering the special status of weapons, the disciplining of individual soldiers into cleanly working parts, and the crucial role militarization has played in the fostering of industrialization and automation it is clear that war has been a major force in the drive to integrate humans and machines into effective complex systems. The culmination of this process was World War II with the genesis of computers and the elaboration of incredibly complicated human-machine systems: ships, fleets, planes, wings, weapon teams, armies.

In the confusion of postmodern war's identity crisis, the military has looked to technology for a way out. Dreams of smart weapons and bloodless infowar are just that, so instead the cyborg soldier has been created. But cyborg soldiers die like normal humans so the basic problem remains. Besides, even cyborg weapons systems can go horribly wrong, as my case study of the destruction of Iranian Flight 655 by the *USS Vincennes* demonstrates. This tragedy took place because the US Navy was using a state-of-the-art computer system, the Aegis, to "manage" their incursion into Iranian territorial waters. But where the humans and the computer system interfaced there was failure. Fear (always present in battle), faith (in the Aegis system), and scenario fulfillment led to the death of hundreds of civilians. There are no technological solutions to the problem of war, not even cyborgization.

And there are other unintended consequences besides operational failures in thinking so. One of my specific interests has been examining how technology is restructuring gender roles in the contemporary military, creating a series of masculine and feminized roles that have more to do with rank and mission (especially in terms of combat) than sexual characteristics.

The same could be said for exploration in space and the deep sea. There is no function for sex or gender in these settings. The humans are part of a system, with specific roles. Their moments of autonomy are sharply prescribed by ground control, their mission, their space ship, and by space itself.

The folks at NASA are fond of saying that space is just another place to work. Well, that's an exaggeration. Outer space (and aquaspace) are particularly unforgiving environments where you must be a cyborg, and very careful, to survive. But there is some truth in the jest, for most corporate leaders would also argue that you must integrate workers and machines into complex computerized systems effectively if you are going to last long in capitalist-space.

My own experiences as a consultant for Hewlett-Packard were an eye opener. When I first went into the cavernous office space of their Corvallis, Oregon inkjet factory it seemed that the ubiquitous computers were the main focus, with the humans flitting about like bees bringing them honey in the form of data. In my two years working there I saw first hand how crucial effective human-machine communication is in the hyper-competitive world of high tech industry, where H-P, incidentally, excels. It's not that the human is neglected. H-P treats its workers very well indeed. But they also treat their computers well and they pay special attention to how well humans and computers get along.

Industry has led the way in modifying other creatures to be cyborgs, whether it is roaches to explore pipes, mice for genetic research or cloned sheep for the mass production of meat, wool, even human organs. And industry has done incredible things in with modeling (the latest Boeing aircraft was never a physical prototype) and in tracking workers. But in most areas of cyborg research governments are ahead. Yet, the private sector is rapidly catching up, urged on by the unforgiving invisible hand of competition.

Cyborgs have directly paid off in one industry already: entertainment. Mass media, from action figures to movies, is full of cyborgs. A day spent watching children's television will turn up dozens. Talking dinosaurs with implants, mutant turtles, detectives with pop-up heads, and scores of space warriors are among the many varied cartoon cyborgs. Robocop, the Terminators, Luke, Darth, and

CP30 of *Star Wars*, and the many cyborgs of *Star Trek* have been among the most influential cyborgs of the movies. But if you watch the mass media much you'll find you can't escape cyborgs and their issues in any genre. Lawyer and doctor TV shows are full of cyborg plots around transplants, definitions of death, and other real technoscientific issues. Even sports shows and events lavish attention on elaborate equipment, sculptured (with machine work-outs) bodies, and complex reconstructive surgery. Documentary channels and news shows plot the cyborgization of war and industry and the spread of the internet and on the *X-Files* and the SF channel the wilder speculations about cyborgs are explored in chilling detail.

Yet some of the most startling uses of the cyborg come from the art world. Numerous paintings and multimedia displays are on cyborg themes, but it is in performance art, where the artists cyborg themselves, that the limits are being pushed furthest. Stelarc, an Australian, has gone from piercings and hanging himself to the use of elaborate prosthetics as he explores the limits of his flesh, as measured against his desire to be a cyborg. Orlan, a French artist, has undergone a series of cosmetic surgeries, documenting every detail. She has tuned one of the most common cyborg interventions into a performance that raises profound questions about the role of fashion in cyborg medicine. Which is fitting, for medicine is where some of the more startling implications of cyborgization are just becoming clear.

For example, advances in medical cyborg research are redefining the line between humans and animals through xenoplants. All transplants involve cyborg medicine, since the organ, the recipient, and the whole process are managed through systematic literal and conceptual mechanization. Already pig tissue is used successfully for a number of transplants such as cartilage and corneas. Unsuccessful liver xenoplants from baboons have been making the headlines for a number of years and soon, judging by the successes in controlling immune responses, they will be successful. An artificial liver invented in Japan is a cyborg machine in itself since it requires two dogs as an integral part of the apparatus. Liver functions are so complicated we haven't been able to totally simulate them.

Human transplants, and the use of artificial organs and body parts, continues to increase. Modifying ourselves through medicine is becoming more and more common. Now literally millions of interventions are performed to suck out fat or put it in, carve better facial features, modify the immune system, or otherwise "improve" the natural body.

But perhaps the most significant area of medical cyborg interven-tion is in the last years or days of life, when many people are kept alive because they are linked to complex machines. Not only has this become the major cost center in our very expensive medical system but it is also completely changing the meaning of death and life. Working doctors and medical technologists no longer speak of death plain and simple. Patients are "single-dead," "double-dead," or "triple-dead," depending on what machines they are hooked to, what their heart and brain are doing, and whether or not their organs can be harvested. The ability of cyborg medicine to prolong life has been balanced by the systematic prolongation of death for many people, leading to a movement for a right to die.

The rights of the living dead are negotiated in the courts, the legislatures and the hospitals, from the cold precise work of the "procurement specialists" who harvest transplant organs from neomort humans and sacrificial animals to that Johnny Appleseed of Death, Dr. Kevorkian, who has assisted in the suicides of over thirty terminally ill people. Dead mothers birth live children and dearly departed dads father fresh babies. More and more patients live on as brain-dead bodies or in the senile dementia of high tech prolonged dying. Both longevity and euthanasia increase as do political struggles around death. As one newspaper headline put it: "Deciding when you want to die becomes a part of American life." Today the very definition of life and death is a difficult, inexorable, and intimately cyborgian issue.

And so is reproduction. Complex medical interventions have been developed to treat infertility that involve extracting human eggs and fertilizing them in "washed" sperm before reimplantation. Lately scientists have even found a way to do away with the sperm altogether through cloning techniques. So we have the possibility of a 65 year old woman giving birth to her genetic twin. Or, actually, with a few modifications, a 65 year old man could give birth to his genetic twin.

So it turns out that the definition of sex is also an issue of cyborg technology. Of course, transsexual surgery and hormonal therapy can reconfigure a person from male to female, or the other way around. But even on a more subtle level cyborg thinking has exploded our old two-gender system. As infomedicine has replaced biomedicine the body is now conceived of a set of data, including hormone levels and exposures, bodily sexual characteristics, sexual attractions, and genetic markers. Scientists are now arguing that there are actually three, five, even eleven genders depending on what criteria one chooses.

Gender identification it turns out is even more plastic than sex, as any visit to cyberspace will show. Cyberspace is actually a new place and a cyborg place. When you go there you leave your body behind.

CYBERSPACE

Although I'd been around computers for years, working on Unix and MS-DOS systems, my first real experience with cyberspace was organizing an academic meeting there, on cyborgs ironically enough. It was a strange, disembodied, experience. Now, there are those who would say describing an academic meeting as disembodied is a tautology. But rhetoric is one thing, actually meeting in real time without your bodies is reality now and it is certainly different from being there in person, maybe even better. Now at your typical academic meeting the body grows lethargic as it drags from one panel discussion to another until finally even massive transfusions of espresso drinks are incapable of raising its heart rate. The only sure-fire stimulant in such meetings, and probably for academics in general, is talking. Take a comatose, nearly dead, Professor and give him or her the floor and a Lazarus-like resurrection takes place.

Maybe this explains the exhilaration of virtual meetings. First, you can "talk" a great deal, although the "talk" is actually typing messages onto a computer screen. In what truly must seem a miracle (the immaculate conversation?) everyone can talk at once because your text eventually comes up with everyone else's on the screen. So, as you sit there with your mind racing and your fingers dancing the most your body can do is squirm. After several hours of this you want to stand up, wave your arms, and scream.

This is certainly not advisable if you are in a monster computer lab surrounded by a hundred undergrads who are already looking at you askance because of your forty-something appearance and the fact that you are chuckling maniacally at the clever things you are writing while exclaiming hello outloud when a friend logs on from Puerto Rico, and then gasping when someone you just met licks your face!!! Virtually licks your face, I should clarify.... but it is very upsetting just the same.

Five or six hours of being cyborged on-line had incredible effects on my body. When I'd walk outside into the rain I felt lighter, not all there. And I wasn't because my consciousness was still back in that virtual elsewhere in cyberspace where it had just been bumping clumsily (textually) into the other attendees' projections. Talking at once to the simulations beamed from Australia, England, Germany, Puerto Rico, New York, Boston, Seattle, and San Francisco it became impossible to think of the world as anything other than hanging in space showing one face after another to the Sun. After all, morning in Australia was afternoon in Oregon and late evening in London. It reminded me of the reaction all astronauts get once they soar above the atmosphere. "Hey!" they always exclaim, even if they've promised themselves they wouldn't go gooey about it as all the other space traveling cyborgs do, "It's one world. It's just....hanging there in space." Strangely enough cyberspace and outerspace impose the same perception, although in other respects they are so different. Disembodiment in cyberspace is hyperembodiment in outer space, but both places are dependent on machines, and therefore both places are only inhabited by machines, and cyborgs of course.

There is a great deal of fascinating research on cyberculture these days. Sherry Turkle has shown how computers, and particularly the internet, are evocative technologies, bringing out latent habits of mind and body. Turkle also explores how computers mediate communication between people in surprising way, such as the tendency of some humans to prefer confiding in a nonjudgemental machine.

Then others have speculated on the political implications of cyberspace. There is much talk these days of netizens, cyberdemocracy, and a digital nation. It is a bit overblown.

Some partisans of cyberspace forget that they have left real bodies and the real world behind. But as Sandy Stone, cybertheorist and transsexual activist has argued:

> Cyberspace developers foresee a time when they will be able to forget about the body. But it is important to remember that virtual community originates in, and must return to, the physical. No refigured virtual body, no matter how beautiful, will slow the death of a cyberpunk with Aids. Even in the age of the technosocial subject, life is lived through bodies.

While there are political mobilizations in cyberspace, such as organizing for the Czechoslovakian Velvet Revolution, the Zapatista support movement (which triggered the infowar theories of RAND and the Pentagon), coordination of the protests against the Yugoslav government in 1996-97, and the many campaigns for privacy on the Net daily, politics are still decided in the flesh. It's just that much of that flesh is cyborged these days.

CYBORG POLITICS

> The cyborg is our ontology; it gives us our politics. The cyborg is a condensed image of both imagination and material reality, the two joined centers structuring any possibility of historical transformation.
>
> -- Donna Haraway

Public policy is being profoundly impacted by cyborgian technologies in ways the current debates about the information highway, genetic research, high tech military interventions, and doctor-assisted suicides merely hint at. Without a broad, historically rich and philosophically deep understanding of the interrelationship of these issues—in the concept of the cyborg and of cyborg society—these discussions will be impoverished, and, inevitably, ineffectual. This is why I have made the political implications of cyborgization the major focus of my current research and the subject of my next book, *Cyborg Citizen*, due out in 1999.

Conscious political responses to the increasing cyborgization of the human range from the joyous or accepting to the horrified. Unconscious reactions seem to cover the same spectrum as can be seen in cyborg portrayals in the mass media where countless "bad" cyborgs and "good" cyborgs are featured in various cartoons, films and TV shows. Often the same story will have both. But not always.

In 1995, Christopher Reeve, the actor made famous portraying Superman in the movies, fell from a horse and became a quadriplegic. A sad story? No, a heroic cyborg tale. Instead of wallowing in his fate as a barely mobile creature, dependent on and intertwined with machines, trapped in power-beds and wheelchairs, Christopher Reeve has chosen to become a militant advocate for more intimate cyborgization. A quest that within a year of his accident took him to the cover of *Time* magazine and made him a featured speaker at the 1996 Democratic National Convention.

Using his status as the crippled Superman he has unified most of the patient groups in the United States focused on spinal cord injuries. This invalid cyborg front now confidently predicts that science will master the reconstruction of the spinal cord within 30 years! Their goal is to accelerate this medical breakthrough by decades. Despite the obvious difficulties in predicting scientific discoveries, their basic claim is reasonable. Judging by other advances in applied cyborgology there is an excellent chance that total restoration can be developed soon after the turn of the millennium, *if* enough resources are mobilized. Indeed, in August of 1996 researchers announced that they had successfully repaired severely severed spines in mice.

Cyborgs, such as Reeve clearly have their own priorities. As cyborgs proliferate in type and number cyborgian issues will play an increasingly important, eventually fundamental, role in politics. Culture itself, as a whole, will be a cyborg culture, if it isn't already now. However, not all cyborg transformations are as positive as the crippled Superman's current crusade. If anything, there are as many frightening cyborg futures as idyllic.

While many subcultures obsessed with cyborg transitions are individualistic or even libertarian, like the Extropian techno-optimists based in Los Angeles, there are certainly some that are authoritarian.

Consider Aum, the Japanese cult whose ideology was shaped equally by science fiction (Isaac Asimov's especially) and Buddhism. World domination was their goal, nothing less. Aum devotees wore special six volt electrode shock caps (four volts for children) that were meant to synchronize the wearer's brain to their guru's brain waves. These Perfect Salvation Initiation machines cost initiates about $7,000 dollars a month to use. Aum's security/medical team treated dissenters with electroshock and psychoactive drugs. If necessary they were executed and then microwaved into ash, literally. Cult members were told that they were superhumans capable of resisting nuclear blasts and plasma rays, thanks to cyborg technologies (such as the shock caps and drugs) and meditation. Aum's many young scientists not only manufactured small arms but also produced a wide range of biological and chemical weapons, including the sarin used in the Tokyo subway attacks that killed 12 and injured thousands. They were seeking to buy nuclear weapons and develop laser and microwave weapons as well. Fortunately, this particular cyborg microculture collapsed into self-destructive paranoia before it could effectively incorporate mass death weapons.

Subgroups who see themselves as human-machines have a long history, especially in war. The right-wing World War I German veterans who formed the Freikorps, the shock troops of Fascism, explicitly embraced a self-conception of body self-loathing and killing machine worship. Ernst Jünger, their leading poet, rhapsodized about the imaginary man whose "instinctual energies have been smoothly and frictionlessly transformed into functions of his steel body." While today's military explicitly ignores this transformation of instinct into technologized body in popular culture it is the latest thing, in the form of tattoos, brands, piercings and ornamental prosthetics.

Not that such modifications are apolitical. In the piercing, tattooing, and other body modification subcultures there are certainly extreme right-wing tendencies but most such enthusiasts see their art as a glorification of the body instead of a loathing of it as exemplified in the Freikorps early in the century and the Christian Right of today. Politically, cyborgification refuses simple mappings as is clear when one looks at the strange geography of one type of cyborg modifications—implants.

There is a great deal of emotion around the idea of implants, much of it paranoia. The television series, the X-Files has a number of episodes about aliens implanting small objects in human abductees. The paranoia is this series in palpable, and one of its main attractions, yet one doesn't have to look far to see this paranoia in action. The Militia subculture in the US has a great deal of cyber-paranoia, especially about possible government implants.

Considering the reality of the Aum experiments and the years of government research into brain control technology, there are certainly grounds for concern. Throw in hacker legends, military global positioning satellite technology, medical bioapparatus innovations, the relentless miniaturization of electronic technology and the deployment of controlling implants seems almost inevitable.

But it is not. The Leviathan of the state, modern or postmodern, is indeed made up of citizens. While cyborg technoscience may offer states, and other institutions (businesses, religions) tremendous new powers it also offers citizens a chance to communicate among ourselves. The thing is to take it seriously.

> There are several consequences to taking seriously the imagery of cyborgs as other than our enemies. Our bodies, ourselves; bodies are maps of power and identity. Cyborgs are no exceptions. A cyborg body is not innocent;

it was not born in a garden; it does not seek unitary
identity and so generate antagonistic dualisms without end
(or until the world ends); it takes irony for granted. One
is too few, and two is only one possibility. Intense
pleasure in skill, machine skill, ceases to be a sin, but an
aspect of embodiment. The machine is not an it to be
animated, worshiped and dominated. The machine is us,
our processes, an aspect of our embodiment. We can be
responsible for machines; they do not dominate or threaten
us. We are responsible for boundaries. We are they.

-- Donna Haraway

As a philosopher I've tried to do my part for setting boundaries.
This is why I've recently put forward a Cyborg Bill of Rights to get
people thinking about how technoscience is changing what we
consider essential rights, as well as how it might threaten our old
freedoms as well. One takes a risk getting up in front of an audience
of academics and calling for a Cyborg Bill of Rights, as I did at the
University of Illinois in 1997. And I did get a fair number of laughs,
not always where I expected them. But this is part of the philosoph-
er-as-gadfly job, putting forward crazy ideas that might only be
half-crazy.

In all seriousness citizenship is being reconfigured by the cyborg
technosciences and by the direct challenges to the very idea of the
human, which are coming from genetic, medical, computer, and
military technologies. If citizenship is worth preserving it must be
cyborged, with a full understanding of all the relevant technical,
philosophical, historical, and political implications.

One big problem is our limited technical understanding of systems.
It is only now being addressed. The mathematician Norbert Wiener
coined the term cybernetics in the late 1940s to describe the study of
information in systems, both natural and artificial. He made some
key contributions to it, but it is a field that is still in its infancy.

CYBERNETICS

Philosophically it is interesting to think of reality as being made up
of three basic elements: energy, matter, and information. We happen
to know a tremendous amount about matter and energy. Albert
Einstein even showed us that energy and matter can actually be
translated into each other, a discovery that has had its practical
drawbacks. But information is very much a mystery. When people
talk about information theory they usually mean the mathematics to
calculate how much redundancy a signal needs to make sure it is
communicated. There is nothing about what information the signal
carries, and even less about what knowledge is.

In the 1950s computer scientists set out to create artificial intelli-
gences. They were sure it wouldn't be that hard and they made

many confident predictions. Now their predictions are in ruins and their greatest accomplishment has been to show just how complicated the idea of knowing information really is. Still, some progress has been made in the last 3,000 years, roughly the time humans have been keeping track of ourselves.

Our biggest step was probably our first, language. If Noam Chomsky is right, language is hard-wired into our brains. Without language we wouldn't be human. Language is also the net we use to hold information and it is the main tool we use to work it. Logic, the formal relationship of hunks of information to each other, is sometimes helped by language, sometimes not. Finally, mathematics, a strange proto-language all its own, was created out of language, logic, and the fact that much of reality comes in discrete chunks.

All three of these fascinating forms of information, language, logic, and mathematics, have been improved over time. But we have also come to understand their limits. The ambiguities and lapses of language are well known to any speaker and the paradoxes of logic have been paraded for the amusement of the general populace by smart-ass philosophers since ancient times. But it wasn't until Hans Gödel took an old philosophical paradox and mathematicized it that it was provable that mathematics wasn't the language of God; instead it is a formal system that had to be flawed by being either incomplete or with at least one paradox.

This is actually one of the key insights of the fledging field of cybernetics: all formal systems are limited. Alonzo Church and Alan Turing showed that Gödel's incompleteness theorem also applies to infinite computers. So even before it was well underway thoughtful technoscientists knew that computerization had its limits. Gregory Bateson was one of a number of cyberneticists who put forward another limit on systems: a system cannot understand itself. When you throw in the insights from quantum physics that the observer effects (becomes part of) the system they observe and that often to know one thing (like the position of a reached electron) precludes knowing something else, you've pretty much reached the negative boundaries on postmodern epistemology.

But there is a positive aspect to cybernetics, starting with Wiener's main fascination, feedback. He chose the term cybernetics because it was based on the ancient Greek word for steersman. Feedback is what keeps a steersman on course, taking in the information from the wind, sea, and boat and using it to turn the rudder this way and that. Feedback can be positive (carrot) or negative (stick) but in most complex systems it is both. Wiener also noticed that feedback and other system dynamics were often basically the same in both artificial and natural systems. This is what makes the cyborg possible after all, communication across the divide between the living and the inanimate.

In the last few decades some amazing discoveries have been made. Complexity theory, often misnamed as chaos theory, has shown how many complex systems that seem chaotic are really following complex and/or counter-intuitive patterns. Some systems are so unstable at certain points that a very small input can change them radically. This is known as the butterfly effect, after the example that a butterfly in Brazil could change the weather in Montana. True enough, but don't make the common mistake of thinking that every Brazilian butterfly is so influential. It is only in very unique, singular, cases that such a small effect can have such large impacts. And it is because systems tend to stabilize. Sometimes it is the stability of entropy, the much feared heat-death of the universe, but sometimes it is the stability of a higher complexity, as with the dissipative systems that the chemist Illya Prigogine described mathematically which won him the Nobel Prize. Prigogine's math works best on phase changes such as the transition of a supersaturated solution into a crystal with just one grain of salt, but he has also argued that life itself is a dissipative system.

Gregory Bateson used to say it is the pattern "of patterns that connect." Fractals, where patterns reproduce themselves on different scales, as with snowflakes and landscapes, are a good example of this. The patterns, rules if you will, of systems apply to cyborgs in my opinion.

Bateson also said,

> We might regard patterning or predicability as the very essence and raison d'être of communication... communication is the creation of redundancy or patterning.

This relationship of patterns to communication is a crucial one, for communication and literature both. Which is maybe one of its best investigators in the English professor, N. Katherine Hayles. She has shown how discourse itself is a cybernetic system, one we can modify by talking or writing about it.

Another valuable approach is to start with communication theory and work back to history and meaning. This is what Marshall McLuhan's last student, Paul Levinson, does in his natural history of the information revolution.

So it isn't just technological breakthroughs that are driving the information revolution, the revolution in systems thinking has played a key role. Some of the best theorists have returned to Wiener's argument that artificial and natural systems are basically the same. Kevin Kelly, for example, has written a book called *Out of Control: The New Biology of Machines, Social Systems, and the Economic World* that links them together along with the libertarian political agenda that is championed by *Wired Magazine*, which Kelly happens to be executive editor of. "Out of control" doesn't mean running amok. It means outside of external control; these systems run on their own dynamic. They can't be directed. By looking at a lot of contemporary systems research, on everything from living coral reefs through new management theory to the building/evolving of little mechanical creatures that are "fast, cheap, and out of control," Kelly has come up with some new system rules, which he calls "The Nine Laws of God."

They are:

- distribute being,
- control from the bottom up,
- cultivate increasing returns,
- grow by chunking,
- maximize the fringes,
- honor your errors,
- pursue no optima—have multiple goals,
- seek persistent disequilibrium,
- change changes itself.

Now, one can argue with this list but there is something to these rules, and to the other insights of systems thinking I've just dashed madly through. And their similarity to postmodern cliches and to the dynamics of cyborgs are more than suggestive. So one of the main tasks of someone who wants to understand cyborgs is using, and helping with, our growing understanding of the dynamics of information itself. Bringing it all together—the historical and philosophical understanding of how new technosciences and intellectual paradigms impact society, the technical details from the main sites where cyborgs are made, and the principles of systems—is what cyborgology is for.

THE NEED FOR CYBORGOLOGY

In the summer of 1995 I visited MIT's media lab and met a couple of grad students working on wearable computers and sophisticated human-machine interfaces who happily labeled themselves cyborgs. Steve Mann was connected to his computer through satellite signals and next to his head antenna he wore a camera that constantly broadcast images on two tiny TV screens he wore as glasses. He could set his camera to show everything upside down or sideways, which he had done to see how long it took his brain to adapt. Or he could set his camera for infrared and "see" the electrical cords in the walls, and even trace the power lines to hidden cameras in Harvard Square shops. Since he broadcast his camera output onto his web page this was something the shops didn't appreciate so he was banned from several of them. Steve called his existence mediated reality, because everything he saw was mediated through his camera.

His colleague, Thad Starner, was working on augmented reality. He had a small laser that painted a computer screen onto one of his retinas. His other eye observed the physical world. He controlled his computer, which he wore, through a one-handed key pad. Through an antenna he also had access to the internet. Most of the time he existed in cyberspace and Cambridge, MA at the same time with his senses simultaneously accessing both worlds.

Within a few years these two cyborgs had multiplied into a small band of graduate students with various types of constantly improving wearable computers. Research into improved interfaces, and power technologies such as tennis shoes that generated electricity when you walked in them, is ongoing. For many young people, being 'borged is empowering.

These cyborgs are also clearly cyborgologists, working away consciously at expanding the possibilities for human-machine integration. But many other scientists and engineers are also cyborgologists, whether they admit it or not: computer designers working on interfaces, surgeons on transplant teams, bioengineers improving all sorts of tools and machines. There is another group of cyborgologists who seek to understand the social and philosophical implications, and possibilities, of our cyborgization. Some claim the label cyborgologist proudly, such as the historians of technology whose special interest group on computers called itself The Cyborgs for awhile, or the anthropologists who issued a manifesto for a Cyborg Anthropology.

Other academics have begun trying to put these changes into perspective without using the image of the cyborg. For example, the historian David Channel sees our current culture as a merging of the longstanding Western ideals of organic order ("The Great Chain of Being") and artificial rationality ("The Clockwork Universe"). He argues that today these two trends come together in the idea of the vital machine. Bruce Mazlish, of MIT's Science and Technology Studies program, tells a slightly different story. His is an epic about the human quest to gradually transcend our illusions. It starts with the rejection of the pretense that we are at the center of the cosmos (the Copernican Revolution), then the illusion that we are fundamentally distinct from animals died (overthrown by evolutionary theory), followed by our realization that we are not even completely rational (thanks to Freud and the unconscious). Finally, the "Fourth Discontinuity," as he calls it in a book of the same name, will have to go. It is the artificial divide we've drawn between humans and machines.

Inevitably people began to see the earth itself as a cyborg system. As Donna Haraway put it referring to the Gaia theory that the biosphere is a self-regulating system, Gaia is a "cyborg world." Considering the domination of humans and our technologies in the biosphere this seems inarguable. Gregory Stock, a physicist and science writer, has given this insight a masculinist twist in his book *Metaman*. He postulates that the earth is one cyborg creature with its own needs and desires, including the procreation of other metamen throughout the galaxy. As cartoonish as it sounds, when you combine the dynamics of the Gaia theory with Stock's impressive documentation it almost becomes a convincing story. Something has happened. Whatever you call it the living system we are part of is clearly both organic and machinic. And it is evolving.

In their 1960 article Clynes and Kline went on after describing the cyborg idea to discuss its implications, the foremost being participatory evolution. Clearly, if humans are modifying ourselves to live in space and other strange places, the dynamics of natural evolution have been supplanted, at least temporarily, by artificial evolution. And artificial evolution isn't just the conscious breeding of farm animals that Darwin discussed, it now includes the direct modification of human bodies and genes. Our interventions are now crude, but the new technosciences that are making the nanotechnology revolution a reality, including in particular genetic engineering, promise that very soon, in terms of natural evolution, we will be creating creatures that can't even be classified as humans.

The probability of post-human cyborgs is one that horrifies some people, and thrills others. We see the beginning of these divisions with the debates on cloning. The Catholic Church declares that cloned humans won't have souls while scientists sign petitions calling for more research. There are even groups like the Extropians who label themselves "transhumanists" and who hail the approaching proliferation of human-based creatures as inevitable and wonderful.

These simple dichotomies are not adequate. First, there are many different types of cyborgs and many different ways to categorize them. Cyborgs can be restorative, normalizing, reconfiguring, and enhancing. People are used to thinking of cyborgs as restoring lost functions, maybe even to the extent that someone can seem "normal" again. But the possibility of reconfigured humans and enhanced humans, really not necessarily humans at all, is one that officially has been ignored. Only in 1997 did the Human Genome project fund research on the use of genetic engineering to enhance and reconfigure humans. Other schemes have been proposed that focus on the systems level of the cyborg (postulating meta and semi cyborgs for example) or that look at the relative balance between biological and machinic elements.

Other scholars, such as Che Sandoval, Joseba Gabilondo, and Cynthia Fuchs have focused on how cyborgs make race and nationality profoundly ambiguous. If identity is plastic on the physiological level, it is certainly so in culture. They argue there is a liberatory potential in the cyborg borderlands and shifting identities.

The second problem with dichotomies is that they may not be the most important ones to focus on. Steven Mentor, Heidi Figueroa--Sarriera and I commented on this in our essay "Cyborgology" in the introduction to *The Cyborg Handbook*:

...perhaps there is as much hope as horror in the realization that China Mountain Zhang comes to in Maureen McHugh's story, "Soon, perhaps, it will be impossible to tell where human ends and machines begin." There are, after all, more important distinctions to make, between just and unjust, between sustaining and destroying, between stable and erratic, between pleasure and pain, between knowledge and ignorance, between effective and ineffectual, between beauty and ugliness.

We go on to elaborate on this.

All of these are dangerous dualities, to be sure, but spectrums we have to face in any event, even if only implicitly or by omission. Once, most people thought that artificial-natural, human-machine, organic and constructed, were dualities just as central to living, but the figure of the cyborg has revealed that it isn't so. And perhaps this will cast some light on the general permanence and importance of these dualities. After all the cyborg lives only through the symbiosis of ostensible opposites always in tension.

We end with a call that our readers consider going beyond dualistic epistemologies and consider the epistemology of cyborgs:

- THESIS, ANTITHESIS, SYNTHESIS, PROSTHESIS, and again.

Reality is not a simple swinging to and fro, nor is it a straightforward march to completion. We are not determined by our technology. We do not construct the world socially. Our technologies, our cultures, our will, and nature all dance together weaving a future from the present. Reality is dynamic, and lumpy. Some things follow from others, some persevere, others seem to just appear. But that is because we can't comprehend everything that is going on. We can understand a great deal. but not everything and any epistemology that pretends we can know it all is seriously flawed. And we've got to get on with figuring out what is happening to us, because for good or ill (probably for both) the era of posthuman possibilities is beginning. To deny it is dangerous. To recognize it is to begin to understand, perhaps even control, our future.

REFERENCES

Channell, David. *The Vital Machine: A Study of Technology and Organic Life*, Oxford University Press, 1991.

Fox, Renee and Judith Swazey. *Spare Parts: Organ Replacement in American Society*, Oxford University Press, 1992.

Gray, Chris Hables. *Postmodern War: The New Politics of Conflict*, Guilford/ Routledge, 1997.

Gray, Chris Hables with Steven Mentor and Heidi Figueroa- -Sarriera, eds. *The Cyborg Handbook*, Gray, Routledge, 1995.

Haraway, Donna. *Simians, Cyborgs and Women*, Routledge, 1989.

Kelly, Kevin. *Out of Control: The New Biology of Machines, Social Systems, and the Economic World*, Addison-Wesley, 1995.

Levidow, Les and Kevin Robins, eds. *Cyborg Worlds: The Military Information Society*, Free Association/Columbia University Press, 1989.

Levinson, Paul. *The Soft Edge: A Natural History and Future of the Information Revolution*, Routledge, 1997.

Mazlish, Bruce. *The Fourth Discontinuity: The Co-Evolution of Humans and Machines*, Yale University Press, 1993.

Stock, Gregory, *Metaman: The Merging of Humans and Machines into a Global Superorganism*, Simon and Schuster, 1993.

THE SPECTRE OF EMERGING AND RE-EMERGING INFECTION EPIDEMICS

by

Donald B. Louria

In the field of infectious diseases one of topics creating the greatest current interest is the potential for future epidemics due to emerging and re-emerging infections. Emerging infections are those that are completely new (such as HIV-AIDS or Ebola virus) or infections that are not new but occur in new geographic areas, or infections that suddenly increase in frequency. For example dengue, a mosquito-borne viral disease that causes an influenza-like illness (but can be fatal) is both increasing in frequency and appearing in new geographic areas. Re-emerging infections are those that reappear after a variable period of dormancy when they were absent or at very low frequency; thus in the last five years there have been more than 100,000 cases of diphtheria in the ex-Soviet Union. Prior to that time, in the 1970s and 1980s there were very few such cases. Another example would be cholera epidemics that suddenly break out in refugee camps. Also included in emerging infections are old, established infections that have recently become highly resistant to the antibiotics we have available to treat them; for example strains of tuberculosis have arisen that are extraordinarily resistant to drugs that used to be very effective.

Epidemics of infectious diseases have always played a major role in the course of human history, altering economic and community stability, determining where people will live, and human migration patterns. In 1918 influenza savaged the planet, killing an estimated 20 million persons, 1 of every 100 persons in the world population. In previous centuries, epidemics of plague and cholera shattered and, not infrequently, decimated communities.

Now we have HIV-AIDS which is in its own way every bit as venomous as the influenza plague and cholera epidemics of past decades and centuries. Since it first appeared in the 1980s the virus has infected more than 40 million people and killed at least 11 million. In some countries in Africa 20% to 25% of the adult population are HIV-positive, a terrifying figure. And it is far from done.

HIV is just one example. The next century will bring with it a variety of other severe infections with the ability to affect large numbers of people, potentially with very high death rates. Faced with the threat of these very dangerous infections, our goals are to detect them early, precisely define the microorganism, intervene

Donald B. Louria is *professor and chairman at the Department of Preventive Medicine, UMDNJ-NJ Medical School, Newark, New Jersey.*

promptly, control them and above all prevent them. I have summarized the approaches to prevention and control in Table I.

TABLE 1

Approaches to Control and Prevention of Emerging and Re-Emerging Infections
• Global surveillance • Improved diagnostic facilities • Improved public health infrastructure • Immunization • Increased research funding • Mitigation of societal determinants

The current emphasis is on worldwide surveillance to allow very early detection. That certainly is necessary. Recently there was a substantial delay in establishing that the Ebola virus infection in Africa and plague in India were causing epidemics. The Centers for Disease Control in Atlanta and the World Health Organization are leading the effort to improve surveillance in every country so emerging and re-emerging infections will be suspected early, proper specimens collected and rapidly transported to sophisticated diagnostic laboratories. A marked improvement in surveillance capabilities requires money, trained personnel and organization. Fortunately, the amount of money needed to establish a more effective (but not a perfect) surveillance system is modest, perhaps a few hundred million dollars worldwide. Not a great amount of money. The major problem in creating effective world-wide surveillance for emerging infections relates to organizational capabilities in less affluent countries. That is a continuing and vexing problem that guarantees that surveillance can be improved but will still be far from optimal. The extent to which surveillance can be upgraded is dependent on the strength of a country's public health infrastructure. In most developing countries the public health infrastructure ranges from inadequate to horrendous. That is why 1.5 billion people do not have safe water to drink and explains in large part the billions of episodes of diarrhea that account for millions of deaths each year. Surely the most basic functions of a country's public health infrastructure are to provide safe water and, uncontaminated food. Only after these basic functions are in place and operating reasonably effectively can a country allocate the resources needed to include proper infection surveillance as part of its program.

Unlike the modest cost of upgrading surveillance to a reasonable standard of proficiency, creating and maintaining an adequate public health infrastructure (which includes doctors, public health workers, hospitals and other treatment facilities) is quite expensive. The problem is that most developing countries spend less than 30 dollars (often a lot less) per person per year on their populations for health. If that is all the money that is available, most of it will be spent on administration and on treatment of established illness, not on the public health infrastructure or surveillance. In the United States annual spending on health by the government exceeds 1,000 dollars per person and we spend huge amounts on public health infrastructure. Even here our public health infrastructure can break down; for example the massive epidemic in Milwaukee involving 400,000 people infected by an intestinal parasite called cryptosporidium was caused by a breakdown in a single water treatment facility (1).

So we can, and will, have better surveillance, but such surveillance cannot be optimal until each country has a very good public health infrastructure—and that is unlikely to happen in the absence of major economic gains among the less affluent countries of the world in Africa, Asia, and Latin America.

A third mechanism for coping with emerging and re-emerging infections is immunization. Millions of children die every year of vaccine preventable diseases, mostly in developing countries, just because they have not been given available safe and effective vaccines. In other cases, as with influenza virus, we know a lot about the infectious agent and we do have effective vaccines, but every few years the influenza virus changes (mutates) rendering previous vaccines ineffective. Although that creates the potential for massive, severe epidemics with a large number of deaths, our scientific community is always prepared, and we develop new and effective vaccines rapidly. In contrast to established infections such as influenza, epidemics of new emerging infections often present a daunting challenge and it may take a long time before an effective vaccine is developed. HIV/AIDS is a good example; it has been with us for almost 20 years and we still do not have a vaccine. Similarly we do not have vaccines for Ebola or Hanta viruses. The bottom line is that immunization is important but we cannot depend on it to prevent epidemics of emerging and re-emerging infections.

In actuality I believe that potentially the most effective approach we could take to ameliorate the problem of emerging and re-emerging infections is modification of the societal variables that provide the mileau in which emerging and re-emerging infections arise and thrive. Since modifying societal determinants can avoid emerging infections entirely, it is true prevention, unlike global surveillance which focuses on early detection, after the epidemic has started.

In Table II I have listed some of these societal variables whose modification or amelioration I believe to be crucial to prevention and control of these infections. The two overarching determinants are population pressures and global warming.

TABLE 2

Major Societal Determinants of Emerging and Re-emerging Infections
<ul><li>Population Growth</li><li>Global Climate Change</li><li>Poverty</li><li>Malnutrition</li><li>Unsafe water</li><li>Dam building and irrigation</li><li>Increasing international and in-country travel</li><li>Public health infrastructure or policy breakdown</li><li>Population displacement and migration</li><li>Urbanization</li><li>Human behavior, particularly as related to sexually transmitted diseases</li><li>Warfare</li><li>Contamination of environment with wastes</li></ul>

POPULATION GROWTH AND ITS CONSEQUENCES

Modern civilization dates from approximately 10,000 B.C. It took almost that entire span for the planetary population to reach one billion persons. Once that milestone was achieved (in 1830) the population express gathered steam. It required only 100 years for world population to double to 2 billion and 70 more years to reach 6 billion. Fertility rates are falling so that some developed countries are at zero population growth but overall world population is growing at a rate of 1.5% per annum with a projected population doubling time of 47 years. It seems almost certain there will be 8 billion people on the planet by the year 2030.

A huge problem is the geographic disparities in moving towards population stability. In North America the population is projected to double in about 120 years. That is quite reasonable. Europe is even closer to zero population growth. In contrast, in Asia (excluding China) the population will double in 39 years. In Latin America the doubling time is 37 years and in Africa it is an almost unbelievable 26 years.

Estimates of the ultimate population of the planet depend on assumptions about changing fertility rates (the average number of children a woman will bear). Very optimistic projections for eventual final population of planet earth are 6 to 8 billion; in contrast very pessimistic estimates are 18 to more than 20 billion. Medium estimates are in the 11 billion range. Actually, as of this writing, there is encouraging news about population growth; the latest United Nations projection suggests that the world population will eventually stabilize at 10.7 billion - almost two for every one person living on earth now. That is a huge increase, another five billion persons, but it is almost one billion less than the 11.6 billion projection by the same group a few years ago.

Whatever projection is used it seems clear that the crowding that characterizes much of the current human condition will increase—and crowding promotes the spread of newly emerging and re-emerging infectious diseases. Some of the adverse effects that may occur consequent to a doubling or tripling of world population are summarized in Table III.

TABLE 3

Potential Consequences of World Population Growth to More Than 10 Billion People That Will Increase the Risk of Emerging and Re-Emerging Infections
• Increased global warming • Increased potential for person to person disease spread • Rainforest destruction • Wetland destruction • Larger numbers of travelers • Increase in wars within or between nations • Increased number of refugees and internally displaced persons • Increased hunger and malnutrition* • More crowding in urban slums • Increased numbers of people living in poverty • Inadequate potable water supply* • Ever more large dam construction and irrigation projects * New technologies could prevent or minimize

I have included rainforest and wetland destruction. These do not usually directly impact on emerging infections but they may. As flies, mosquitoes and rodents are deprived of their usual ecological niche in forests or wetlands, they may migrate to population centers, become established and spread a variety of infections. I have put an asterisk by hunger and inadequate potable water. Hunger is associated with malnutrition and malnutrition predisposes individuals to all sorts of infections, but it may be that with newer technologies we can feed a population of 10 or 15 or even 20 billion people.

Similarly our technologic proficiency may allow us to transform ocean water into drinking water. That is not particularly likely, but theoretically we could have enough potable water for 15 to 20 billion persons.

One of the most important consequences of the massive increase in population numbers has been the progressive urbanization of the planet. The latter half of the 20th century has been characterized by creation of large cities with ever greater population densities that draw inhabitants from rural areas and often act as sink holes for hordes of newly landless people displaced by economic hardships or the ravages of war. Lured to the cities by hopes of a better life, large numbers of migrants experience underemployment or no employment and become inhabitants of teeming slums. The crowded slums, often with inadequate sanitation and stagnant water, are not only ideal settings for the initiation of epidemics and then rapid person to person spread but are also virtual breeding grounds for rodents and for fly and mosquito vectors. At the beginning of the 20th century about 15 % of the world population of less than 2 billion people lived in cities; at the beginning of the 21st century approximately half of the world's 6 billion persons will be urban dwellers; that figure is expected to grow to 60 to 65% by the year 2030; 80% of this massive urbanization will take place in the so-called developing countries. At the dawn of the 21st century there will be at least be 26 megacities, each with more than 10 million inhabitants.

Doubling of our present population absolutely guarantees a planet with a very large number of crowded cities, each with populations of one to 30 million persons, most with areas of teeming slums. And that in turn translates into a profound increase in the risk of contagious diseases, some of which will be emerging and re-emerging infections and some of which will turn into rapidly spreading, severe epidemics. The only way to avoid this is to limit population growth. Ten or 11 billion people on the planet will be bad enough; 14 to 18 billion would be an infectious diseases catastrophe.

Perhaps no human activity is so conducive to emergence or re-emergence of infectious diseases as warfare, a human activity that, measured by the number of people involved, becomes more extensive every century (in part because of population growth). The 20th century has been the bloodiest in history. There have been 150 wars in the last half of the century, resulting in more than 23 million deaths, 2/3 of them civilians (2).

Wars create the milieu for infection in many ways: massive injuries that invite microbial infection; forced migration of non-immunized persons into areas inhabited by disease-carrying vectors (mosquitoes, flies); crowding in refugee camps with inadequate sanitation facilities; exposure to disease-carrying rodents; malnutrition, even starvation; and destruction of public health infrastructures and safe water supplies. Additionally, mass rape as an accepted or deliberately overlooked behavior of conquerors, or sometimes as an intentional military and governmental technique of intimidation and cruelty, can become a vehicle for spread of sexually transmitted diseases.

The prospects for less warfare are not good. The ethnic, religious, racial and tribal strife that currently savages planet Earth will be worsened by population growth, crowding and competition for increasingly depleted natural resources. In future decades competition for a decreasing supply of fresh water will be an ever more important motivation for combat. Water shortages will result from a combination of population growth, increased irrigation require ments to feed that growing population and the effects of urbanization and industrialization.

As the planetary population increases, the need to produce more food and find more sources of energy will result in massive irrigation projects and dam building. In 1950 there were 5,000 dams designated as large (defined as more than 15 meters in height). Now there are 38,000 of these dams, and 1,200 more are under construction; 2/3 of these will be more than 30 meters in height (3). So we are building more and larger dams primarily to create electricity. Additionally, increase in the land under irrigation is cutting into the remaining forests and damaging our crucial wetlands.

In developing countries new infectious disease patterns have often been an untoward consequence of vector redistribution resulting from dam construction. Habitat advantages have been created particularly for mosquitoes and snails (4). The Aswan dam resulted in 200,000 new cases of a serious virus infection 'Rift Valley' Fever as well as a marked increase in the parasitic disease schistosomiasis (from snails) and malaria (from mosquitoes) among populations adjacent to the dam. In Ghana construction of a massive dam on the river Volta was followed by a huge increase in schistosomiasis. Additionally, large dams, such as the gargantuan one being constructed on the Yangtze River in China, require relocation of tens of thousands, even millions of people; this forced displacement is often associated with worse living conditions and increased poverty, both risk factors for increased infection rates. In some cases the enormity of the construction project requires the recruiting of large numbers of workers from distant parts of the country or from other countries. These new "temporary" immigrants bring their own infections with them, creating the potential for emerging or re-emerging infections in inadequately protected local populations. Furthermore, the newly created large bodies of water are subject to massive fecal contamination, increasing the likelihood of re-emerging pathogen epidemics (for example cholera).

As the world population increases and travel becomes easier and more affordable, it is inevitable that the number of international travelers will increase. In 1990, there were 280 million persons traveling outside their own nations' borders. By the year 2000 that figure will increase to between 400 million and 600 million people. Some of these travelers will carry organisms with them and start epidemics in countries to which they travel. Others will become infected during their travels and bring the organism back with them to their own countries. And of course infected mosquitoes, flies and rodents love to hitch rides on airplanes or sea-going vessels.

GLOBAL CLIMATE CHANGE

The climatological event most likely to affect infectious diseases is global warming. The evidence for global warming gets progressively more convincing (5). It is likely that in the 21st century the planet will be 1.5 to 4°C warmer. The warming will result in winners and losers; some geographic areas, now quite cold, will be more productive; others will suffer from severe floods or drought and the entire world will experience more severe and turbulent weather events, in particular ferocious storms. At least half the greenhouse (heat trapping) gas burden that is responsible in considerable part for the warming is due to carbon dioxide. Methane (10-20%), chlorofluorocarbons (10-20%) and nitrogen oxides (4-7%) are the other principal greenhouse gasses. More than 25 billion tons of CO_2 are released yearly into the environment. Of that amount, about 60% results from industrial and home fossil fuel use and 10 to 20% from emissions from the more than 500 million cars and trucks used around the globe. An additional 20% comes from felling and burning trees in the world's ever diminishing forests. Atmospheric carbon dioxide concentrations have risen 30% in the last 2 centuries; the current concentration of 360 parts per million may increase by an additional 60% by the end of the 21st century.

Changes in temperature will alter the distribution and behavior of vectors of infection, including mosquitoes and flies. At higher temperatures some mosquitoes tend to be more active, eat more voraciously and bite more frequently. Additionally they have more rapid reproductive cycles and the time required for development of infectious agents, such as malaria, in the mosquito is lessened. Mosquitoes that have found higher elevations cold and inhospitable will be able to thrive in previously mosquito-free areas that become warmer; this in turn will introduce certain mosquito borne diseases, such as malaria and dengue, to unexposed areas and therefore to non-immune populations. This introduction of infectious agents as a result of vector (mosquitoes, flies) redistribution into non-immune populations is likely to be one of the major world-wide consequences of global warming.

Malaria, dengue, and schistosomiasis lead the list of infections likely to increase as a consequence of global warming (6,7). Malaria is perhaps the most feared disease on the planet. There are an estimated 300 to 500 million people affected each year and 1 to 3 million deaths. Only tuberculosis rivals malaria in the number of deaths. Global warming could markedly increase the number of cases. Dengue is a viral disease transmitted by mosquitoes that affects 100 million persons yearly. In most persons it causes fever, chills and severe muscle aches with recovery in a period of several days to several weeks. However, in those who have had previous infections, a second attack can produce severe bleeding and even death.

The future appears bleak in regard to global warming. Carbon dioxide accounts for more than half the global warming (greenhouse) gases. Although the United States is the world's leader in production of greenhouse gases, contributing more than 20% of total annual carbon dioxide burden, China, now in second place, is increasing its CO^2 release at a stunning rate (up 27% in less than 10 years) (8,9). With understandable determination to become more affluent, China will require a profound increase in energy, almost certainly by use of its abundant supply of coal. That presents a gargantuan problem.

But it is not only the progressive increase in carbon dioxide levels that is of concern. Methane could be an equally important greenhouse gas in the next century (10). Methane traps heat far more effectively than CO^2. Its sources are much harder to control. They include: swamps, marshes, wetlands, fens and bogs; coal, oil and gas extraction; rice paddies; termites; and the intestinal tracts of ruminants. The rule of thumb is for every 1 billion new people on planet Earth, there will be 500 million more cattle; the more cattle, the more methane.

Since there will be difficulties in controlling methane emissions, the only solution will be to control population increase and the number of cattle and to compensate for the likely methane increase by a marked reduction in the main greenhouse gas, CO^2.

One important reason population growth and global warming are such critical variables is that once the global population has grown excessively and the planet is warmer, there is no logically thought out behaviors of man that will quickly remedy the situation in regard to these two superordinating determinants. To achieve cooling of a planet heated by human actions would take at least several decades, plenty of time for the planet to be savaged by multiple severe emerging infections. If excessive population growth creates immense problems, it would take centuries to effect significant reductions in that population size.

POVERTY, MALNUTRITION AND EXPOSURE TO UNSAFE WATER

A reasonable estimate is that between 1/4 and 1/3 of the population of planet Earth live in poverty. Some 2 billion persons suffer from undernutrition or malnutrition; 1 billion experience hunger on a daily basis and at least 400 million suffer from dire poverty and major nutritional deprivation. In the range of 1.5 billion people lack access to basic health care. A similar number do not have access to safe water. With a substantial increase in population size expected in the 21st century, these disturbing figures could get substantially worse.

The link between poverty and undernutrition or malnutrition is clear; so is the connection between under-nutrition and infectious diseases. Malnutrition increases the frequency and severity of many infections. Those living in poverty around the planet usually are also the ones particularly exposed to unsafe water. The adverse consequences of that exposure are fully established; diarrheal disease from water-contaminated by bacteria, viruses and parasites accounts for billions of episodes each year and as a consequence billions of dollars in lost productivity, millions of deaths and an extraordinary toll in hardship and misery.

DISPLACED PERSONS

In a hotter world with a larger population, huge numbers of people will be displaced by floods, drought, and reduced ability to produce enough food. At present there are approximately 50 million persons who are either refugees outside their own borders or internally displaced. In the year 2050 with 9 to 10 billion persons on the planet and the uprooting effects of global climate change, there could be hundreds of millions of refugees and internally displaced persons. That is likely to be beyond the coping capacity of man. That huge number of refugees, often living in squalor, crowded together in unhygienic circumstances, creates an ideal situation for development of emerging and re-emerging infections with rapid person-to-person spread and then spread outside the refugee camps to local populations.

Once established, given the ease of travel, the infections can reach virtually any part of our planet.

The determinants I have discussed are, for the most part, intensely interrelated. Thus, as the number of people on Planet Earth increases by billions, the likelihood of global warming increases. These two variables together influence other variables, such as warfare and forced migration, that establish the setting in which emerging and re-emerging infections arise and thrive. Although population size and the extent of planetary warming are the overarching issues, their control will not eliminate wars, poverty, increasing urbanization, increasing energy demands, forest destruction, resource depletion, etc. But a modicum of control can be the difference between manageable problems and uncontrollable problems. That in turn can be the difference between a relatively small number of emerging and re-emerging infections of limited severity and more frequent epidemics of far greater severity.

Recognition of the inter-relatedness of the determinants is important because it leads to a systems (holistic) approach rather than a more simplistic, but more comfortable, focus on single determinants with the implicit (but erroneous) assumption that modification of single lesser variables can have a major impact without a simultaneous change in the entire system, particularly in the two superordinating factors- global warming and population growth.

If these variables, these determinants, can decide the frequency and severity of infection epidemics in future decades and centuries, then if we can modify them we can to some extent be the masters of our fate. There is no way to prevent all major epidemics. No matter what we do some major epidemics of emerging and re-emerging infections will occur and some will kill tens of thousands or hundreds of thousands or even millions of people. But we have the capacity to determine whether there will be many or only a few major epidemics severe enough to kill or damage large numbers of people.

There may be no way to avoid an eventual planetary population of 10 billion, but prudent actions could make sure we do not have a world population of 14 or 18 billion. That would be a huge difference. Similarly we may be committed to a 1^0 to 2^0C increase in global temperature but we should be able to avoid a 4^0C or greater rise.

As we enter the new millennium we desperately need a sense of urgency in approaching the two major determinants, population growth and global warming. Unfortunately that sense of urgency is virtually non-existent. The consequences of insouciance are illustrated well by the following alternative scenarios.

The eventual size of world population is dependent on three variables: births; deaths; and length of life. Medical science, public health and better living conditions act together to reduce death rates. We cannot do anything about that. The variable most within our control is births. The total fertility (the number of children each woman will bear) for the planet is now 3.0. When it gets to 2 children per family on average we will be at replacement fertility and population numbers will stabilize. To many, the difference between a total fertility rate of 3.0 and a rate of 2.0 would seem to be no big deal. In point of fact it is a very substantial difference and reducing the rate to 2 in developing countries will not be easy. An optimistic scenario is that the world could reach replacement fertility of 2.0 by the year 2040. A somewhat less optimistic scenario foresees replacement fertility by the year 2080. That is only a 40 year difference but that 40 year delay in reaching replacement fertility translates to an extra 3.0 to 3.5 billion people on the planet when it achieves population stability (see Table IV). That is the difference between 10 billion people and 13.5 billion people just because of a short delay of a few decades in reaching replacement fertility. That sense of urgency should be heightened by the realization that every 10 year increase in life expectancy adds another 2.5 billion more people to the eventual planetary population. Life expectancy in the world is approaching 70 years. It will reach 80 years sometime in the next century and that means 2.5 billion more of us just because of one more decade of life.

A combination of a modest increase in life span plus a modest delay in reaching replacement fertility means the difference between 10 billion people on this planet and about 16 billion.

If world population grows to 12 to 16 billion and the societal judgment is that this number is too large for the health of the planet,

TABLE 4

Effects on Eventual World Population of Changes in the Life Expectancy and Time of Achieving Population Control (Replacement Fertility)		
Life Expectancy (years)	Replacement Fertility Achieved (year)	Differences in Population at stability
70	2040	2.5 billion
80	2040	
70	2040	>3 billion
70	2080	
70	2040	
90	2080	>8 billion

what in the world could we do to correct the situation? The answer is nothing. It would be too late.

The same concerns about leadership and a sense of urgency apply to global warming. The sense of urgency on the part of our leaders is for the most part rhetorical. The meeting of 100 countries in Kyoto, Japan in December 1997 was an opportunity to chart a firm course for the planet. It did not happen. The United States recognizes the needs to reduce greenhouse gas emissions to at least 1990 levels. We will even agree to cut back to 1985 levels. That sounds good but, in point of fact, we have supposedly been focused on that need since the Rio conference of 1992. What has in fact happened? Since 1990, our greenhouse gas emissions have increased by about 11% and the percentage increase each year is accelerating. By the year 2005 when we say we will start doing something about our situation, it is estimated that we will be 30% or higher in per capita carbon emissions compared to 1990. So, the 11% increase since 1990 will be allowed to grow an additional 20% before we take any significant actions. That represents billions of tons of carbon dioxide and other greenhouse gases that will remain in the atmosphere, heating our Earth for decades.

Once our greenhouse gas emissions have increased by 1/3, cutting back to 1985 or 1990 levels will be very difficult, indeed impossible, in the absence of draconian actions; even then we may not be able to achieve the promised reductions.

How can we best approach these issues? The answer is there is plenty we can do. There are three domains—individual actions, changes in our educational system and political actions.

INDIVIDUAL ACTIONS

1. Accept the premise that the societal variables relating to emerging infections are crucial and therefore efforts to mitigate them are worthwhile.
2. Become involved as activists. That is not difficult. There are a large number of local, state, or national groups and organizations that focus on one or more of the major societal determinants. There is nothing wrong with a primary focus on a single determinant so long as it is done in the context of systems thinking—a recognition of the complexity of the interacting variables and a realization of the impact of one determinant on other determinants.

One of the rationales used for continuing non-involvement is the perception that individuals are feckless and are unable to exert any major influence on events. That is a mistake. As Sidney Smith, a 19th century English clergyman noted (slightly modified), "No man or woman makes a greater mistake than he or she who does nothing because he or she can only do a little."

An example of the potential for individual actions is global climate change. The public views global warming as a distant phenomenon unrelated to their daily lives but in point of fact individuals can have an impact on 2/3 of carbon dioxide emissions. The New Jersey Chapter of the World Future Society has launched a "Think globally, act locally" campaign directed to individuals and will recommend specific actions that individuals can take that will markedly reduce their personal contributions to global warming (Table V). Similar campaigns should be conducted in every state. Obviously impacting global climate change requires the cooperation and involvement of a large number of countries around the world. But the United States has to take the lead and that requires efforts by individuals, industry, legislators and the executive branch of government. If individuals take appropriate actions that in itself might persuade or compel industry and politicians to follow their lead.

EDUCATIONAL MODIFICATION

Much of the vision about educational change in future decades focuses on technological advances that permit ready access to information, distance learning and sharing information and projects through sophisticated communication technologies. For some the emerging technologies also enhance the potential for systems thinking. The extraordinary technological cornucopia has the potential for improving education in both developed and developing areas of the world but in one critical area our educational system is failing young people, and that is in instilling long term commitment

TABLE 5

What Each of Us Can Do to Reduce Global Warming

1. **Buy and drive vehicles wisely**

- Motor vehicles account for 32% of all the CO_2 we spew into the air.
- Think twice before you buy that gas-guzzling sports utility vehicle.
- Look for a car with overdrive that gets 30 miles per gallon or better
- Keep your auto well maintained. Make sure your air conditioner is not leaking.
- Shop locally to cut down driving miles. Stay within the speed limit.
- Walk or bike short distances (it's good for your health, too).
- Use public transportation, car pools and ride in groups whenever possible.

2. **Conserve electricity**

- Buy appliances that have the ENERGY-STAR efficiency label (or its equivalent).
- When replacing light bulbs, buy energy-efficient compact flourescent bulbs; they cost more, but last longer and use only 1/4 of the wattage of a regular bulb. In the end, you save money.
- Turn off lights, TVs, computers and other appliances when not in use.

3. **Conserve heat and hot water**

- Put full loads in dishwashers and washing machines (wash small loads by hand).
- Wash clothes in warm or cold water.
- Use low flow shower heads.
- Keep room temperatures no higher than 68°F in winter and no lower than 74°F in summer.

4. **Some Specific Tips for Home Owners**

- Make sure your home is properly insulated, especially around windows and in the attic where a lot of heat may be lost. Proper attic insulation by itself can make a big difference.
- Get a free energy and insulation evaluation by your utility company.
- Have your furnace and air conditioner serviced regularly.
- Rake leaves instead of using a gas or electric blower.
- Plant trees and flowers that provide shade and beauty, absorb CO_2 and reduce the amount of lawn to be mowed.

to participating in attempts to solve or ameliorate major societal problems at local, state, national or international levels. In the middle of the 20th century some teachers attempted to instill such commitment as part of courses in civics. In the latter part of the century a small number of schools mandated participation (usually ephemeral and superficial) in community projects or community activities. But as we enter the new millennium, commitment to the society, local or global, has for the most part been profoundly subordinated to the glamour of the technologic imperative. That is a grievous mistake. We remain a meliorist society, still able to solve the major problems that face us by the dint of our own efforts. But, in the absence of commitment of young people to solving critical problems, ineluctably one or more of these problems will either become insoluble or controllable only at such a high price to the society that debilitating societal disruptions will inevitably follow. At that point we will no longer be regarded as meliorist and that new perception will have its own unpleasant consequences including an ever increasing momentum towards hedonism with a consequent drug abuse problem that will dwarf anything we have thus far experienced.

How do we avoid that unpleasant scenario? Certainly over the last two decades some futurists have made valiant attempts to achieve commitment by involving students and community members in problem solving activities. Unfortunately these laudable efforts have been limited in extent and have not been widely adopted.

I would suggest we need a small but critical change in our educational system that could have far reaching and beneficial effects.

I would propose that at every educational level students have, on a mandatory basis, interdisciplinary courses (or interdisciplinary curricula) with the following goals:

(1) To provide them with a reasonably comprehensive view of the major issues facing the society at local, national and global levels
(2) To have them understand that many of the major issues are inter-related
(3) To teach the importance of systems (holistic) thinking about critical problems
(4) To have them think like futurists drawing up alternative scenarios for future years or decades and then suggesting ways of moving towards the most attractive of the scenarios
(5) If possible, to improve their communication skills
(6) Above all, to consciously instill in them long term, part time or even full time, commitment to solving or ameliorating one or more of the important problems at local, state, national or international levels.

These courses should probably start at the junior high school level, certainly in high schools and be a part of college and graduate school (law, medicine, business) curricula.

At the New Jersey Medical School, we started a one-month elective for senior students eight years ago titled "Medicine, Society and the

Future: The Doctor as an Activist" that focused on issues at the interface between medicine/science on the one hand and the greater society on the other. The elective was such a stunning success that two years later we decided to offer a small number of sessions in the required first and second year preventive medicine courses. Outside lecturers were selected in part because of expertise in a specific area but also because they were excellent communicators and role models with a history of constructive activism.

Attendance was obligatory but first year students were told there would be no second year sessions unless 80% voted for them. In no year did student approval ratings fall below 80% and 8 to 20% in different years indicated that the small number of sessions had changed their perceptions of medicine and their futures in the profession.

Encouraged by the success of the medical school sessions we are now collaborating with a nearby college and a high school. The plan is to have an interdisciplinary course for college freshman taught by faculty from three universities. At the high school we will "train" the teachers in the hope they will, in turn, incorporate the goals enumerated above.

The college course may be titled "Critical Decisions, Society and the Future: the College Graduate as a Constructive Activist." Students finishing the 28-session course will be allowed to take a for-credit elective consisting of community-based projects, a component of which will be the mentoring of students from the participating high school.

This college high school program is in its inchoate phases. It is a small beginning but hopefully will be successful and then can serve as a prototype for national policy.

If we do not modify our educational offerings to achieve the goals I have enumerated, we will continue to fail our young people and they will not be committed to societal problem solving. That in turn would also mean inadequate attention to the societal determinants that will determine the frequency and severity of emerging infection epidemics in the next century.

POLITICAL ACTIONS

Emerging infections may be dangerous, but they also offer an extraordinary opportunity to use the threat of these infections to compel politicians to face and take action against the major problems facing our society - problems that also happen to be the major determinants of emerging and re-emerging infections. Politicians around the world show a remarkable ability to avoid doing anything aside from rhetorical hand wringing about the issues discussed in this chapter—population explosion, global warming, wars, growing numbers of refugees, massive urbanization with equally massive slums, dwindling supply of fresh water, etc.

Part of the reason for this insouciance is that the problems, important as they are, do not strike a responsive cord among the constituents of those politicians. The problems seem remote and not applicable to the daily lives of most people, in particular people in the more affluent developed countries. After all, who cares if another billion people are added to the world's population so long as that billion is in the developing world and out of sight, and who cares if hundreds of millions of children in those same developing countries suffer from horrendous poverty and hunger, or if another 10 million people are added to the growing rolls of refugees.

It has been said that politicians think of the next election, statesmen of the next generation. We have a lot of politicians but very few statesmen in the so-called developed world. In the United States, the public is interested in the present or in the next few years, in issues such as taxes and crime—not in world population, the greenhouse or refugees in far off countries. Indeed the public apathy in the United States is so pervasive that the majority of adults do not even care enough about their democracy to vote. If people do not know about or care about the critical issues facing the global society, it is hardly surprising the politicians do not feel any pressure to focus on these issues. But the concerns about emerging and re-emerging infections could change all that.

Politicians and their constituents may have a lot of trouble in relating to, or doing anything about, expanding world populations or the threat of global warming, but those same politicians, and their constituents can understand, and be concerned about lethal epidemics of the AIDS virus, or the Hanta virus that kills by overwhelming pneumonia or Ebola virus that kills with bleeding everywhere. For many of these and other killer organisms, we still have no effective treatment, no preventive vaccine.

We can use the specter of, indeed the virtual certainty of, severe epidemics of emerging and re-emerging infections to persuade politicians that insouciance is no longer acceptable and that they must focus upon and take steps to mitigate these major societal determinants. The biggest mistake we can make is to assume that emerging infections can be relegated totally to the domain of scientists and the medical profession. Aside from immunizations, all the medical profession can do is in essence mop up for a society that doesn't have the willpower to take the preventive actions that are needed. The medical profession can lessen severity of most, but not all, epidemics by surveillance and early intervention, but the most critical aspect of emerging infections, prevention, can only be achieved by the actions of the greater society.

We need a combination of individual activism, educational modification and public pressure that modifies the behavior of politicians and societal leaders. One thing is clear. We must have activists who work to change our political behavior so that we tackle the major societal determinants of emerging and re-emerging infections. If we do not succeed as a global society in achieving that objective, the future of our global society will be bleak in regard to savage emerging infection epidemics.

REFERENCES

1. MacKenzie, W.R., N.J. Hoxie, M.E. Proctor, M.S. Gradus, K.A. Blair, D.E. Peterson, J.J. Kazmierczak, D.G. Addiss, K.R. Fox, J.B. Rose, and J.P. Davis. A Massive Outbreak in Milwaukee of Cryptosporidium Infection Transmitted through the Public Water Supply. *New England Journal of Medicine,* 331:161-167, 1994.

2. Sivard, R.L. *World Military and Social Expenditures.* World Priorities Inc., Washington, D.C., 1993

3. Gardner, G., and J. Perry. Dam starts up, p. 124-125. In L. Starke (ed.) *Vital Signs 1995.* Worldwatch Institute, Washington, D.C., 1995

4. Hunter, J.M., L. Rey, K.Y. Chu, E.O. Adekolu-John, and K.E. Mott. *Parasitic Diseases in Water Resources Development.* World Health Organization, Geneva, Switzerland, 1993.

5. Houghton, J.T., L.G. M. Filho, B.A. Callander, N. Harris, A. Kattenberg, and K. Maskeil. Climate Change 1995. The Science of Climate Change. Contributions of Working Group 1 to the Second Assessment Report of the Intergovernmental Panel on Climate Change. Cambridge University Press, Cambridge, United Kingdom, 1996.

6. Patz, J.A., P.R. Epstein, T.A. Burke, and J.M. Balbus. Global Climate Change and Emerging Infectious Diseases. *JAMA,* 275:217-223, 1996.

7. Rogers, D.J., and M.J. Packer. "Vector Borne Diseases, Models and Global Change. *Lancet,* 342:1282-1284, 1993.

8. Flavin, C., and O. Tunali. 1996. "Climate of Hope: New Strategies for Stabilizing the World's Atmosphere". Worldwatch Paper 130. Worldwatch Institute, Washington, D.C., 1996.

9. Flavin, C. Slowing Global Warming: a Worldwide Strategy. Worldwatch Paper 91. Worldwatch Institute, Washington, D.C., 1989

10. Khalil, M.A.K. (ed.). *Atmospheric Methane: Sources, Sinks and Role in Global Change.* Springer-Verlag, Berlin, Germany, 1993.

GENETIC ENGINEERING: OUR NEW GENESIS

by

Clifton E. Anderson

A brand-new beginning for planet Earth may not be in prospect, but altered environments and human-created life forms will be distinctive features of the new millennium. Ready or not, the world is entering the age of genetic engineering. Plants may be transformed into miniature factories producing plastics, medicines or perfumes. Animals with human genes may produce hearts and other organs for critically ill people.

Cloning—the process of making an exact replica of an organism or genetic material—has been accomplished with animals, and human cloning eventually may be performed. Gene therapy research is exploring strategies for dealing with cystic fibrosis, fragile-X syndrome and other devastating genetic diseases. The transfer of genes within and between microbes, plants and animals provides opportunities for altering life forms and even creating new ones.

People are apprehensive concerning genetic engineering. They realize there is genetic information about themselves recorded on their deoxyribonucleic acid (DNA). In fact, their DNA pattern is as distinctive as their fingerprints; it can be used to identify them conclusively. Personal privacy may be compromised by genetic screening—a test that reveals defective genes and the person's susceptibility to inherited diseases. Medical files containing screening results could prejudice people's jobs if leaked to employers.

Genetic engineering will impact the major environmental problems of our time—overpopulation, pollution, erosion and the rapid loss of biodiversity. It is crucially important that we take stock of the new molecular genetics' probable effects on our planet's ecosystems. There continues to be discouraging news on the environmental front. Will the newfound knowledge of genetic engineering produce revolutionary changes in global ecology?

AFTER ONE HUNDRED YEARS

Revolutions are rooted in the past, in the hopes and annoyances of generations of people. Anyone who stands at the frontier of change is familiar with past events and attitudes, and from them he must chart the future. In the 1890s, visionaries could see only the first few steps of the escalator that would carry science to a pre-eminent position of power in the 20th century. The world-changing biological

Clifton Anderson *is professor emeritus at the University of Idaho; editor of* The Horse Interlude *and author of* History of the College of Agriculture at the University of Idaho.

revolution was not yet in view. A popular theme of the late 19th century was *hygiene*—and futurists of that era envisioned the world triumphing over flies, filth, contagious diseases, contaminated meat and milk as well as unsafe processed foods. Not only did they expect great achievements from science, but the hygienists of 1899 also anticipated 20th century programs to promote public health. They assumed that government would need to support medical research, enforce food safety laws, set standards for drugs, improve the adequacy of health care facilities and increase the competence of health care providers.

During the 20th century, the hygiene revolution has succeeded far beyond the expectations of its supporters 100 years ago. Development of vaccines and antibiotics, plus expanded health care and education, gave people longer lives—a mixed blessing in the impoverished Third World countries. Overpopulation is now a critical problem, with Africa and much of Asia failing to increase food resources in proportion to the burgeoning growth of their food consumers.

Experts say the global population explosion is a phenomenon that dwarfs the power of the nuclear bombs that ended World War II. George Moffett (1994) explains: "If an atomic bomb as destructive as the one that destroyed Hiroshima had been dropped every day since August 6, 1945, it would not have stabilized human numbers. Indeed, two bombs per day would not offset today's net daily population increase of 250,000."

Undoubtedly the most serious problem we face today, the mounting population crisis strains natural resources. Approximately 93 million people live in Mexico—and the world's population shows an increase of about 93 million people each year. A conservation expert warns that "any future increase in population can only lead to a reduction in the food and land available to each person" (Dobson 1996, p. 213).

ECOLOGY AND JUSTICE

Many feminist scholars see parallel ideologies shaping gender discrimination and environmental degradation. They say people who acquiese in the subjugation of women are unlikely to maintain carefully and respectfully the web of interrelationships which unifies all sectors of nature into functioning ecological systems and sub-systems. Power and privilege go together—and domineering individuals claim the right to exploit both natural resources and human resources.

In the past, visionaries such as Saint Francis of Assisi and Albert Schweitzer preached reverence for all life. For many world leaders, the quest for power has been more alluring than the pursuit of justice. According to Val Plumwood (1994), an "ideology of the domination of nature pays a key role in structuring all the major forms of oppression in the West....It has supported pervasive human relations of domination within Western society and of colonization between Western society and other societies as well as supporting a colonizing approach towards nonhuman nature."

Domineering individuals in developed countries distrust the "chaotic" and "defiant" realm of nature, Plumwood believes (*ibid*.) Instead of collaborating with natural processes and rhythms, the developers proceed with plans to reform and modify nature. For them, commodification of natural resources is the ultimate goal of development. Their bottom-line concerns are profits and economic growth.

Right now, genetic engineering is in the development stage. Third World leaders are trying to protect scarce genetic resources from Western exploitation. At stake here is control of genetic material from unusual plant and animal species. Western companies want the material for genetic engineering projects and hope to obtain it without undue expense.

Because newly developed agricultural and medical applications of genetic engineering carry big price-tags; only limited Third World use of the new technologies is expected—at least for years to come. As a result, the economic gap between rich and poor nations is likely to widen. While biotechnology is revolutionizing health care, manufacturing and food production in the West, what steps can be taken to help the Third World participate in the revolution and share its benefits?

SUSTAINABILITY FOR SURVIVAL

Traditionally, American farmers who worry about environmental damage have protected their land with cover crops and other anti-erosion measures. Since World War II, chemical pesticides and fertilizers have become pollution threats—and progressive farmers are responding by using non-polluting alternatives. Some farmers also changed their farming practices in order to save fuel and help check the rapid depletion of oil resources. Today, conservation of soil, water and energy are goals of "sustainable" agriculture. Instead of relying on purchased fertilizers and pesticides, conservation-minded farmers work at developing technologies that minimize the need for off-farm production inputs. Government-aided research and education programs encourage farmers to appraise ecological problems on their farms and to seek workable solutions.

Neglecting sustainability, most American farmers pattern their operations on the industrial model. They standardize their farming methods, not treating erosion-prone fields differently from their other land, and they make extensive use of chemical fertilizers, pesticides and fossil fuels. They seek short-term profits and are not very concerned about environmental damage. As genetic engineering develops, a variety of its products may benefit the environmentalists as well as the exploiters. If geneticists originate perennial grain crops that need not be replanted every year, this would help check tillage erosion. Also, it would reduce fuel usage. On the other hand, new crops suitable for intensive, industrial modes of production might invite environmental damage. For example, a chemical company sponsored a genetic engineering project to develop a pesticide-resistant variety of cotton. The company manufactured herbicide and wanted to enable farmers to use more of it on their crops. This plan for increasing the use of herbicide did not interest sustainable farmers.

Genetic engineering will bring far-reaching changes to pharmaceutical manufacturers, food processors and other branches of industry. At present, sustainability is given support by a few industrial leaders, but many do not assign a high priority to environmental protection. In this time of industrial globalization, American firms setting up operations abroad are able to cut expenses by taking advantage of their host countries' lax anti-pollution laws. Their advantage here is a financial one. Environmental protection would make sense to industrial entrepreneurs if they could be sure that environmentally kind methods of production would also rate high in profitability.

Sustainable farming is well suited to the needs of Third World countries. The costs of importing farm machinery, pesticides, fertilizers and tractor fuel are extremely burdensome for farmers there. Sustainable agriculture's goal is low-input production. A farmer who has a great store of knowledge about how to grow good crops with few expensive inputs is the mainstay of any sustainable farming system. This individual is a knowledgeable, efficient manager. It is essential to provide farmers with opportunities to learn management skills that are required in sustainable enterprises. The educational programs developed jointly by American agricultural specialists and experienced farmers could be adapted for use overseas. For farmers who are not adequately trained, a low-input system of production could net disastrously low returns.

Genetic engineering products designed for use by Third World farmers would be helpful. Farmers would welcome high-protein, disease-resistant varieties of wheat, corn, rice, sorghum and cassava. George Moffett reports: "It could be a decade or more before the potential advantages of biotechnology trickle down to the less developed nations. With little profit to be made, little research is likely to be undertaken where it's needed most—at least not without prodding and money from private foundations or public sector donors" (1994).

BEWARE OF RUNAWAY GENES

It's a scenario that could become a nightmare: genetically transformed bacteria are released into fields, perhaps in a fungus control exercise. The bacteria mutate and promptly display wide-ranging lethal powers. They kill any plant, animal or human they encounter. The infestation spreads, turning vast acreages into desolate wasteland. Less spectacular—but also deadly—would be mutant bacteria replacing all beneficial soil-dwelling bacteria—various species that convert waste into rich fertilizer, those that fix nitrogen from the air into forms plants can use and others that perform useful tasks.

Genes from a genetically engineered crop might escape in pollen that could fertilize the crop's wild relative—a weed. Next, we could have the appearance of a super-hardy weed—a hybrid plant that inherited the transformed crop's ability to poison hungry insects and to withstand big doses of weed-killer. Insect-proof, hard-to-kill weeds would not be welcome in 21st century agriculture!

Sometimes, the transforming genes inserted in plants simply will not work as intended. Recently, the seeds for genetically engineered tomatoes and cotton were taken off the market after farmers who grew the new crops reported disappointing results. A tomato that kept its flavor intact after prolonged storage apparently had unexpected defects. The variety of cotton mentioned earlier was resistant to herbicide, but it inexplicably dropped its cotton balls after being sprayed with weed-killer. The result: costly crop failures.

Because there are great uncertainties involved in the testing of biotechnology theories, work on gene therapy for humans is proceeding cautiously. The prospective rewards of gene therapy are tremendous; the goals are prevention of inherited disease. At present, gene therapy is being directed at *somatic cells*—the working cells in a human body that pass on genes to other cells in that body but do not pass on genes to the next generation. Therapy someday will be directed at *germ cells* that do transmit genetic information to the next generation. Such therapy would alter a person's heritable traits—and this could lead to elimination of inherited diseases. However, mistakes in germ-line gene therapy could bring extreme deficiencies and horrendous mutations to future generations. This is an area of medical research in which work must progress with great care. No errors can be tolerated in germ-line genetic therapy.

THE HIGH COSTS OF PERFECTION

In the future, when gene manipulation has become a precise procedure, prospective parents may be faced by a wide range of possibilities. Of course, they will want to make sure that, at their offspring's embryo stage, gene therapists correct any disease problems or tendencies due to defective genes. Parents also may want the therapists to boost their children's intelligence, add inches to their height and endow them with superior athletic ability, curly hair, blue eyes and a complexion of the most fashionable hue. The possibilities for genetic transformation are likely to be vast, but only for wealthy parents. Designer babies would not be affordable for ordinary Americans or the majority of people in other countries. The existence of a special class of persons who, from birth, were assumed to be "superior" would put slashes in the fabric of society.

A century ago, social reformers were claiming the *eugenics movement* could improve the quality and character of American and European populations. The movement's objective was to discourage "inferior" persons from reproducing. Supporters of eugenics cited sociological studies that showed certain families generation after generation begat criminals, paupers and mentally handicapped persons. To stop such families from producing more socially defective children, the eugenicists proposed compulsory sterilization of persons who became public charges. Sterilization laws targeted at habitual criminals and mental defectives were enacted in about two-thirds of all states. Indiana's law, passed in 1907, was the first in a long series. Laws were passed, amended and repealed. By 1985, at least 19 states had compulsory sterilization laws on the books (Reilly 1991).

Is it sound governmental policy to permit some people to reproduce while compelling others to be sterilized? Through the years, state governments in the United States compelled 60,000 American citizens to submit to involuntary sterilization. Many did not know the purpose for the surgery they received. All this was done "in the name of science" (Reilly). Actually, the scientific claims made by the eugenicists were invalid and their purification program reflected ethnic and racial prejudices, according to Allan Chase (1977). He documents a resurgence of eugenic thought in recent academic discussions about "racial intelligence."

The eugenicists' arguments appealed to Adolf Hitler. In 1933, Hitler supported German legislation mandating compulsory sterilization of the mentally ill, alcoholics and epileptics. The Nazi program for ethnic purification later included killing millions of Jews, other minorities and social misfits in the infamous concentration camps. Nazi genocide convincingly discredited ethnic and racist eugenics, but proposals for eliminating genetic problems in humans continue to be heard. The economic argument often is advanced; it is said to be impractcal to continue treating patients with incurable diseases or uncorrectable deficiencies.

If genetic manipulation of designer babies becomes available for well-to-do parents, an identifiable "superior" caste may rise to the pinnacle of our social order. Do we want to have such a caste? What would be the consequences of a society possessing a privileged elite group who receive their status by means of prenatal genetic engineering? These are important questions to ponder before society permits super genes to be awarded to children of the highest bidders.

Genetic therapy would receive public acceptance more readily if the possibility of preferential treatment for the wealthy could be precluded. Philip Kitcher predicts that complaints about gene therapy would be rare indeed "if a future society assures equal access for all its citizens (and) if it attends first to urgent health needs before creating the opportunities to enhance capacities" (1996).

CAN BIODIVERSITY BE SAVED

Nature is never idle. Year by year, decade by decade, nature is testing new genetic models of microbes, plants and animals. Some models survive the tests while others cannot withstand the rigors of the contemporary environment. If the environment changes, new models that possess appropriate genes for survival will replace older, now outmoded models. Warren Pope, a veteran wheat breeder, tells me "nature does most of the work" when new genotypes of wheat undergo testing. If a genotype under study shows great vitality despite harsh weather and heavy disease infestations in nearby plots, Pope says he may have a winner—always pending further exhaustive trials. He says it is also possible that a disease-resistant variety may perform well for a number of years and then fall victim to a new strain of cereal disease. When conditions in wheat fields are changed by farmers planting disease-resistant wheat, disease organisms are challenged to adapt to the new environment. The result: new races of disease that can overcome the defenses of disease-resistant varieties.

Plant breeders recognize the danger of down-sizing the gene pool of any crop species. They hope to maintain genetic defenses the plants may need under future adverse conditions. The problem of down-sizing became apparent when a new race of corn leaf blight unexpectedly swept through corn fields in the United States in 1970. Most hybrid corn breeders were using restricted gene pools, sacrificing genetic diversity for the sake of expediency. Almost all hybrid corn varieties were closely related. When the new race of blight appeared, virtually all hybrid corn varieties were susceptible to the disease. If the corn varieties grown would have diverse lines of heredity, the corn blight infestation might not have assumed epidemic proportions. As products of science not thoroughly checked out by nature, genetically engineered crops might invite devastating crop failures.

In a wild, uncultivated area, a diversity of life forms exists—each of them interacting with others in complex ways. Together, the organisms comprise a functioning ecosystem. People and other ecological hazards can disrupt the natural biodiversity. Hour by hour, minute by minute, nature's inventory of life forms is dwindling. Although this planet has millions of species of living organisms, many are becoming extinct. Tropical forests are the habitat for about half of earth's biological species—and each minute of every day 100 acres of this habitat disappears. At this rate, virtually all of the estimated 2.3 million square miles of tropical forests will be gone by 2040.

For their genetic engineering projects, scientists located valuable genetic material in unlikely places. An unusual bacterium, *Thermus aquatica*, discovered in a hot spring in Yellowstone National Park, played a vital role in the development of a process to synthesize

DNA. The bacterium is able to grow at 86 degrees Centigrade, a temperature at which other bacteria are killed. A heat-resistant enzyme produced by the bacterium has made possible an important genetic engineering procedure, the polymerase chain reaction technique.

On our planet, individual species evolve and many eventually perish. Life goes on, however, and genetic diversity generates new species and provides support and sustenance to existing species. When humans fail to correct problems that are crippling an ecosystem, many organisms are in peril. "Most of the world's political, economic and health problems are intimately linked to the way we manage the world's impressive variety of wildlife and natural resources," says Andrew Dobson (1996). He continues: "It is not too late to save a large amount of the earth's remaining biodiversity but the time is running very short."

Genes from exotic plants, animals and bacteria are involved in many successful gene transfers. From a purely selfish standpoint, humans should see that protection of genetic diversity around the world is in our interest. After all, genetic material from a remote location someday might provide life-saving therapy for ourselves or members of our families. Protecting a particular species would not be practical; species exist in ecosystems—and survival depends on the preservation of threatened ecosystems. For genetic diversity to be saved, it is essential that people develop greater ecological awareness and a willingness to reappraise human relationships with the non-human world.

ANIMAL-HUMAN PARTNERSHIPS

Medical researchers employ animal gene therapy in order to develop and refine new approaches to the correction and prevention of human health problems. A transgenic mouse possessing human genes—the first animal to be patented—is being used as a model for humans in ongoing cancer research. Other animals are altered so that their bodies may become factories for producing biochemicals for pharmaceutical purposes. The drug components are for human use and are collected from animals in blood samples or in milk and urine. Scientists also consider animals promising sources of organs for transplantation into people. Hearts from transgenic pigs would contain human genes and therefore might not be recognized—and rejected—by human immune systems as being foreign tissue.

Insidious disease organisms that endanger people—including the virulent form of influenza that caused heavy casualties in 1919—appear to find refuge inside animals and may remain there for long periods of time. Concern over infection risks may delay the widespread utilization of animal organs for human transplantation. The idea of using animals for people's medical needs has wide acceptance, however.

At present, the public generally take a utilitarian view of animals' role in food production. The overcrowded conditions in which poultry and veal calves are housed do not arouse much protestation. In an earlier era, cattle received names and continuing individual attention; in today's large-scale dairies, cows are milked and fed by workers who follow split-second schedules. The cows spend their days standing or lying on concrete slabs—surfaces that shorten a cow's productive life by inducing bone deterioration and lameness.

It is difficult to focus on the living landscape as though it were ecology-based and not centered around the needs and wants of *homo sapiens*. If it is true that humankind's survival on this planet depends on our working to protect environmental systems and sub-systems, we must try to understand that everything is connected to everything else in the all-encompassing web of life. We may progress to the point where we look upon nature as "a process, one that is more powerful and longer-lasting than human societies and human beings" (Merchant 1996). To begin with, we at least should think earnestly about our relationships with domestic animals. Is it possible to achieve mutually beneficial animal-human partnerships? Will we not find it diffucult to address issues concerning insects, plants, rodents, wild game and micro-organisms if we are uncertain about our relationships with domesticated animals whose lives interact with our own?

LOOKING FORWARD TO 2099

The number of living species in the world will continue to decline in the next century. There will be less genetic diversity—and, subsequently, total global resources will be diminished. By contributing to erosion, pollution and additional processes of environmental damage, people have set the stage for a catastrophe of major proportions. Dinosaurs and other now-extinct species fell victims to occurrences with which they could not cope. Our turn may be approaching. Peter Ward warns:

"Those who do not believe that mass extinctions are enormous catastrophies would do well to study the past." (1994). Some commentators expect genetic engineering to help future generations stretch their limited resources. It is even possible, suggest Reiss and Straughan, that the risks of human germ-line therapy may have to be endured so that a sizeable portion of the human population may continue to survive on this planet despite "some future, unforeseen crisis" (1996). The prospect of the environment deteriorating to the extent that people will need genetic altering in order to survive is a chilling idea. Instead of relying on genetic engineering, it should be preferable for people to take action now and prevent future environmental disasters.

For many people, environmental protection is an obligation they can fulfill by recycling bottles and newspapers. Others see greater challenges and more extensive obligations. People interested in religion or ethics are conscious of a need to serve others and work for a better future. Similar, but with altered emphasis, are the views of humanists who have begun to view the world from a comprehensive, holistic perspective. Even though their basic approaches to ecology differ, people are becoming aware of their need to communicate with each other and to cooperate in setting priorities for environmental action.

The era of modernity has ended, we are told, and we have entered the post-modern age. Post-modernist thought emphasizes people's present-day freedom to disregard conventional mindsets and leap over the traditional divisions of academic disciplines. The important thing now is to frame ideas and propositions clearly and to participate in ongoing exchanges of opinion. It's an exciting time to be alive, says Walter Truett Anderson (1995). He offers a helpful suggestion for dealing with the great diversity of opinion among today's thinkers: note that only a few persons speak your language, while others are holding forth in their own special languages. Learn to be multilingual, Anderson advises. He explains: "The functioning person in the post-modern world needs to be able to think rationally and understand science, able to appreciate and draw on a social heritage, and able as well to drink from the well of ecological and spiritual feeling that is being tapped by neo-romanticism" (p. 116).

Traditional conservationists and the young firebrands who teach "deep ecology" need to understand each other and try to agree on a few items they can include on an agenda for joint action.

Environmental action must include a variety of public interest groups, with links between them being forged at the grass-roots level. Through cooperation, people with a variety of viewpoints can sort through genetic engineering issues—plus other ecological concerns —and coordinate their efforts to influence governmental policy. The task before us is well-described by Russo and Cove (1998) in the preface of their carefully researched book:

- Science itself is neither good nor bad. It is the uses to which it is put that raise ethical questions.
- Genetic engineering is only a technique.
- We have to decide if we want to use it, and when and how.
- These decisions should be the responsibility of all. We hope that all responsible citizens will wish to take an active part in devising the ethical issues created by genetic engineering.
- Only by being informed, can these decisions be made democratically.

The concerns people are expressing about genetic engineering could mark the beginning of effective citizen action regarding global ecological issues. The longer we delay acting, the less likely are our prospects for saving ecosystems and the biodiversity they represent. Decisions we make now will be important, culturally and politically. At the end of the century, we stand at the frontier of opportunity, creating a new genesis.

REFERENCES

Anderson, Walter Truett. *The Truth about the Truth: De-confusing and Re-constructing the Postmodern World.* New York: G.P. Putnam's Sons, 1995.

Chase, Allan. *The Legacy of Malthus: The Social Costs of the New Scientific Racism.* New York: Alfred A. Knopf, 1977.

Dobson, Andrew. *Conservation and Biodiversity.* New York:Scientific American Library, 1996.

Kitcher, Philip. *The Lives to Come: The Genetic Revolution and Human Possibilities.* New York: Simon & Schuster, 1996.

Merchant, Carolyn. *Earthcare.* New York: Routledge, 1996.

Moffett, George D. *Critical Masses: The Global Population Challenge.* New York: Viking Press, 1994.

Plumwood, Val. "The Ecopolitics Debate and the Politics of Nature" in *Ecological Feminism.* Karen Warren & Barbara Wells-Howe (eds). New York: Routledge, 1994.

Reilly, Philip R. *The Surgical Solution: A History of Involuntary Sterilization in the United States.* Baltimore: The John Hopkins University Press, 1991.

Reiss, Michael J. and Roger Straughan. *Improving Nature? The Science and Ethics of Genetic Engineering.* Cambridge GB: Cambridge University Press, 1996.

Russo, Enzo and David Cove. *Genetic Engineering: Dreams and Nightmares.* New York: Oxford University Press, 1998.

Ward, Peter. *The End of Evolution: On Mass Extinctions and the Preservation of Biodiversity.* New York: Bantam Books, 1994.

MAKING CONTACT: THE MOST IMPORTANT EVENT OF THE NEXT 1000 YEARS?

by

Allen Tough

INTRODUCTION

Between now and the year 3000, what event will produce the most powerful impact on human society?

The answer to this question depends on whether the most powerful event of the next 1000 years turns out to be negative or positive.

If the highest-impact event is negative, it will most likely be a war with worldwide consequences, possibly the end of our culture or even the extinction of humanity. That is why it is so important for us to eradicate warfare before the weapons become even more powerfully destructive. In this paper, that is all I will say about the gloomy topic of future warfare, except to urge you to read elsewhere about the future of warfare and how we could adopt peaceful solutions to our conflicts.

This paper focuses on a happier scenario. What if the highest-impact event in the coming millennium turns out to be a positive event? What will it be? Most likely it will be contact with other civilizations in our galaxy.

INTELLIGENT LIFE

In recent years, scientists and the general public have realized that intelligent life may well be found throughout the universe. It is extremely unlikely that we are the only civilization in our galaxy; it may even contain dozens or hundreds of civilizations scattered among its 400,000,000,000 stars.

When we receive richly detailed messages from some of these civilizations or when their superintelligent probes reach our planet, the effects on human civilization will be profound and pervasive. Few events in the entire sweep of human history will be as significant and far-reaching, affecting our deepest beliefs about the nature of the universe, our place in it, and what lies ahead for human civilization. Seeking contact and preparing for successful interaction should be one of the top priorities on our civilization's current agenda.

Allen Tough, *author of* Crucial Questions About the Future, *is professor emeritus Futures Studies at the University of Toronto.*

CORE VALUES

Many astronomers, biologists, philosophers, and others now believe that the existence of diverse life throughout the universe is a supreme value.[1] That is, in the entire universe, nothing is of greater value, importance, or significance than advanced civilizations and intelligent species—including our own, of course. If asked, "What thing or idea is more important or valuable than diverse life throughout the universe, including human civilization?" many people would reply, "Nothing; human and other intelligent life is the most important thing in the universe."

Perhaps a similar answer will be given by human beings 1000 years from now, especially if interaction with advanced extraterrestrials has occurred by then. Advanced extraterrestrials themselves might also give a similar answer.

Let me emphasize that *human* civilization, at present and in the future, is a significant part of all life in the universe. For us, long-continuing human life is a supreme value of ultimate importance. Because we do not yet, as far as we know, have contact with extraterrestrial life, our top priority at present must be our own civilization. At the same time, however, we should continue and enlarge our present efforts to make successful contact with intelligent life from some other part of this galaxy.

Perhaps some sort of grand project is under way to spread highly positive life (marked by love, compassion, cooperation, wisdom, intelligence, knowledge, harmony, and effectiveness) throughout the universe. We cannot contribute much at present to the flourishing of extraterrestrial species but we can choose a flourishing future for our own human species as one of our fundamental priorities. As Carl Sagan concluded in his Cosmos television series, "Our obligation to survive and flourish is owed not just to ourselves but also to that Cosmos, ancient and vast, from which we spring."

AGE AND CAPACITIES

Any other civilizations in our galaxy are probably much older than human civilization.

Two factors support this assumption. First, the vast majority of stars in our galaxy are much older than our sun, many of them millions of years older. It follows, then, that any civilizations on planets revolving around those stars likely arose much earlier than our own civilization did. Second, it seems quite possible that some civilizations survive for a million years or even longer. If the civilizations in our galaxy range in age from a few thousand years up to a million years old, then we are one of the youngest: by most definitions, human civilization is not much more than 10,000 years old.

Because other civilizations in our galaxy are thousands of years older than human civilization, they have probably advanced in certain ways beyond our present level of development. Some civilizations presumably fail to survive once they discover nuclear weapons or other means of extinction, but surely others learn to cope successfully with this problem and then survive for a very long time. Some of them may be 100,000 years or even millions of years more advanced than we are.[2]

THE LONG-TERM CONSEQUENCES OF CONTACT

An encyclopedic radio message, interaction with the intelligent computer in an automated probe, or some other forms of contact with another civilization will likely be highly significant events in our future. These contacts will likely affect our civilization profoundly at the time and for several centuries afterwards. Indeed, as we look ahead at the long-term future of human civilization, we realize that one of the highest-impact events of all time will probably be contact with another civilization.

What will the specific effects of extraterrestrial messages, interaction, or intervention turn out to be? Four sorts of long-term consequences are particularly likely to result: (1) practical information; (2) new insights about certain major questions; (3) a transformation in our view of ourselves and our place in the universe; (4) participation in a joint galactic project. Let us examine each of these in more detail.

> 1. We might well receive practical information and advice that helps our human civilization to survive and flourish. Possible examples include technology, transportation, a new form of energy, a new way of producing food or nourishing ourselves, the importance of halting population growth, more effective governance and social organization, fresh views on values and ethics, inspiration to shift direction dramatically in order to achieve a reasonably positive future. The message might also bring home to people the importance of eliminating warfare or at least eliminating weapons of extraordinary destruction. Viewing ourselves from an extraterrestrial perspective might be very useful in alleviating our civilization's problems.

Such deep-seated changes will no doubt produce enormous disruption, at least for a short time. We might suffer from massive culture shock and temporarily feel inferior or lose our confidence in our own civilization. Disruption could also occur in the sciences, in business and industry if we learn about new processes and products, in the legal system if we move toward cosmic or universal laws, and in the armed forces and their suppliers if we eliminate the threat of war. Such disruption will probably be tolerable and short-lived. It is best regarded as simply the major cost we have to pay for incorporating new knowledge and possibilities.

> 2. We might gain new insights, understanding, and knowledge about major questions that go far beyond ordinary practical day-to-day matters. Topics in an encyclopedic message could include astrophysics, the origin and evolution of the universe, religious questions, the meaning and purpose of life. The message could include detailed information about the sending civilization, which might be deeply alien to us, and about its philosophies and beliefs. Similar information could be provided about several other civilizations throughout our galaxy, too.

We might even receive a body of knowledge accumulated over the past billion years through contributions by dozens of alien civilizations throughout the galaxy. "Included in this vast body of knowledge, something we might call the 'galactic heritage,' could be the entire natural and social histories of numerous species and planets. Also included, perhaps, would be extensive astrophysical data that extend back countless millennia, providing accurate insights into the origin and destiny of the Universe."[3]

> 3. A richly detailed message from an alien civilization might transform our view of ourselves and our place in the universe—even our ultimate destination. We might gain a much deeper sense of ourselves as part of intelligent life and evolving culture throughout the universe.

Michael Michaud pointed out that "contact would be immensely broadening and deprovincializing. It would be a quantum jump in our awareness of things outside ourselves. It would change our criteria of what matters. We would have to think in interstellar, even galactic frames of reference.... We would leave the era of Earth history, and enter an era of cosmic history. By implying cosmic future, contact might suggest a more hopeful view of the universe and our fate."[4]

> 4. We might eventually play a role in some joint galactic project in art, science, philosophy, or philanthropy. Such a project might aim to solve fundamental mysteries of the universe, help other

civilizations develop and flourish, or spread harmonious intelligent life throughout the galaxy. We could participate in this project through two-way radio messages, despite the length of time required for round-trip communication.

Angelo has noted that contact "might lead to the development of branches of art and science that simply cannot be undertaken by just one planetary civilization but rather require joint, multiple-civilization participation across interstellar distances.... Perhaps the very survival and salvation of the human race depends on finding ourselves cast in a larger cosmic role—a role far greater in significance than any human can now imagine."[5]

It is also possible that our culture will be overwhelmed by an advanced alien culture. However, all terrestrial examples of contact between two cultures have involved physical contact rather than radio messages. Also, terrestrial contact has usually involved territorial expansion by the stronger culture. "If contact has occurred without aggression, the lesser culture has often survived and even prospered."[6] We might well adopt portions of the alien culture but avoid being completely overwhelmed by it.

REFLECTIONS

During the past few years, the scientific search for extraterrestrial intelligence has been growing rapidly and has become quite mainstream within science. It seems reasonably likely that our first contact with another civilization will occur within the next few decades. This first contact will, in turn, lead on to redoubled efforts and then to several additional contacts during the coming millennium.

Of all the positive events that occur during the next thousand years, surely this interaction with other civilizations will have the most profound and pervasive impact on human civilization.

REFERENCES

1. Allen Tough, "From which aspects of reality can we gain meaning and purpose?" Chapter 8 of *Crucial Questions about the Future* (Lanham, MD: University Press of America, 1991 and London: Adamantine, 1995). See http://members.aol.com/AllenTough/mp.html.

2. Allen Tough, "Seven search strategies." http://members.aol.com/AllenTough/mel.html

3. Joseph A. Angelo, Jr. *The extraterrestrial encyclopedia: Our search for life in outer space.* New York: Facts on File, 1985. p. 23.

4. M. A. G. Michaud. "The Consequences of Contact". *AIAA Student Journal*, Winter 1977-1978, p. 20.

5. Angelo, p. 23.

6. Angelo, p. 27

THE FUTURE OF GOD

by

Robert B. Mellert

> The insane man jumped into their midst and transfixed
> them with his glances. "Were is God gone?" he called
> out."I mean to tell you! We have killed him,—you and I!
> We are all his murderers! . . . Do we not hear the noise of
> the grave-diggers who are burying God? Do we not smell
> the divine putrefaction?—for even Gods putrefy! God is
> dead! God remains dead! And we have killed him!" . . .
> Here the madman was silent and looked again at his
> hearers; they also were silent and looked at him in sur-
> prise. At last he threw his lantern on the ground, so that
> it broke in pieces and was extinguished. "I come too
> early," he then said. "I am not yet at the right time. This
> prodigious event is still on its way and is travelling,—it
> has not yet reached men's ears."[1]
>
> —Nietzsche, The Gay Science

Friedrich Nietzsche, the illustrious German philosopher and author
of the above soliloquy, wrote these words in 1882.

He participated in a wave of atheism that spread through the
intellectual circles of Europe in the late nineteenth and early
twentieth centuries, but that never spread popularly through the
masses. The idea of the divine demise, however, did not itself die;
a movement by (curiously!) theologians resurrected Nietzsche's thesis
in the 1960's, amidst the other forms of radical thinking that
characterized that decade. The cover of *Time* magazine for April 8,
1966, summarized it best with the bold-faced headline, "Is God
Dead?"

But just as for Nietzsche, this publication of God's obituary seemed
once again to be premature. Despite the theologians' questions, the
next few decades marked a rise of religious fundamentalism among
many Christians and Muslims, and a return to traditionalist thinking
among many Jews. Today, 96% of the American population say they
believe in God, a slight increase compared with surveys done half a
century earlier.[2] If he were to appear today, Nietzsche's madman
would still find that he had come too early.

What is the future of God? This is the title of the essay, but
perhaps it is not the best way to frame the question. The problem is
the word "God." It may seem like a simple word, but "God" doesn't
mean the same thing to everybody: various images of the deity ap-

---

Robert B. Mellert *is a professor of philosophy, Brookdale Community College,
Lincroft, New Jersey.*

pear throughout history and in different religious cultures. So the first issue we need to look at is a semantic one. We need to study the way in which people have understood the God in whom they believe, and what they believe today. Then we can address the real topic of this essay: What concept of God is emerging for future believers?

We can begin, of course, the way many futurists begin, by looking at the past and observing the relevant trends. Fortunately, this project has recently been done by Karen Armstrong in her book *A History of God*, and by Jack Miles in *God: A Biography*. Anyone interested in a deeper analysis of this aspect of the subject can consult these insightful volumes.

One common interpretation of the origin of the Western image of a single, distinct God is that He arises out of a more ancient era of polytheism. Indeed, the first books of the Bible themselves tell us how the Israelite God Yahweh forbids his people to bow down before other gods, suggesting the existence of parallel deities. In many cultures today God is not singular, but part of a spiritual tribe of deities with individual tasks and separate followings. Hindus, for example, have never found reason to abandon their pantheon. While this may seem primitive to Westerners, who have been reared with the idea that there can be only one God, polytheism does in fact have certain advantages and may not be merely a less sophisticated predecessor of monotheism.

For one thing, if there are many gods, it may be easier to find one whose job description best conforms to what you need to have done. If you are an artist or an expectant mother, you might be able to seek the assistance of a god specially attuned to your situation and more comforting to you than a god who controls the weather (who might be favored by farmers). More importantly, having a variety of gods who specialize in different aspects of life relieves the single great deity of attending to a multitude of very specific concerns. This is simply the application of the economic principle of the division of labor to the realm of religion.

In addition, polytheism creates more confidence for the petitioner: one is more likely to get an answer from a god with an interest and expertise in one's problem, than to persuade the great God to become interested in one's trivial concern. In Roman Catholicism, praying to saints for their intercessory power is a way to save this advantage without compromising monotheism.

The problem with polytheism, however, is that the gods who are interested in specific human concerns generally begin to look and act all too humanly themselves. It requires no stretch of the imagination to think of them as engaging in the same kinds of self-interested behavior, bickering and arguing over jurisdiction, illicit love affairs, etc., as we find among humans. In other words, to the extent that the gods appeal to us with their human qualities, they become less divine and less worthy of worship. To the extent that they are subsumed into a belief in one, great, almighty God, this appeal is lessened.

By contrast, the problem with monotheism is that God becomes so great and so incomprehensible that He effectively ceases to be available for ordinary human concerns. Thus the great trade-off: A God who is truly God (in the Western sense) isn't of much practical use; a god who is one of many scheming, self-interested gods doesn't inspire much worship.

In this essay we will concern ourselves with the Western God of the Jewish-Christian-Islamic tradition. The image of God portrayed in this tradition is, unfortunately, illustrative of the core difficulty in monotheism. This great philosophical conundrum has been called by theologians over the centuries the theodicy problem. It is formulated as a trilemma and can best be illustrated this way:

> Among the following three statements, it is logically possible to reconcile any two of them, but the acceptance of two logically implies the rejection of the third. The three statements are these:
>
> 1. God is omnipotent.
> 2. God loves us.
> 3. Evil exists.

In the first instance, if God is omnipotent and can do anything (create the universe, for example), and if the universe contains natural and moral evils (hurricanes and Hitlers, for example), then it would seem that God lacks compassion for the victims, especially when these victims are innocent sufferers. Dostoyevsky's character Ivan in *Brothers Karamazov* makes this case eloquently when he relates to his brother the story of the innocent children tortured by cruel soldiers in the presence of their parents. How could a loving God allow this?

The second case is the acceptance of a loving, concerned God and the existence of evil in the world. This seems to imply that even God cannot somehow find a way to eliminate the evil, or at least reserve it only as punishment for those who deserve it. A God who cannot do this is a God who must be less than omnipotent. Such a conclusion can give us an image of God as compassionate and loving, but not simultaneously as all-powerful. In this view, there are things that even God cannot control.

Finally, one can attempt to reconcile the omnipotence of God with his loving compassion, but then some explanation is required for the evil and suffering we see around us. It may be that suffering is a test from God (but surely an omnipotent God would know the outcome of such a test before it were given), or that suffering is used by God for greater purposes (but then God's methods would seem either malevolent or inadequate). Or perhaps there really is no evil.

We think of suffering and death, especially of the innocent, as evil, but in the larger scheme of things, these are really good things that we cannot understand. God's ways are not our ways, says the Bible. The omnipotent, all-merciful God created the world this way because it is the best possible of all creatable worlds. (This is the theme that Voltaire develops in *Candide*, where Pangloss rationalizes his way to such a conclusion, whatever the situation.) Since the existence of evil seems so empirically evident, the acceptance of this last approach requires a great faith.

Can the trilemma be resolved by finding a way to reconcile the all-powerful God with the all-loving God without denying the reality of evil? Is it possible to incorporate consistently all three statements into a coherent concept of God?

Attempts to accomplish this task have occupied philosophers and theologians ever since the emergence of monotheism, but always with questionable success. The proposed solutions advocated by each epoch of thinkers are generally intertwined with the common philosophical assumptions accepted in their culture, but these assumptions may not necessarily be accepted in ours. The ancient Zoroastrians, for example, introduced the notion of Satanlike spirits emanating from the one great God, and these spirits are responsible for evil in the world. Somehow, for them, this compromises neither God's essential goodness nor his omnipotence. A similar myth found widely in the Western tradition is that only good spirits were created by God, but some of these freely chose evil, the metaphysical origin of which is not addressed. Later on, the Aristotelian foundations for medieval thought provided a solution for Maimonides and Aquinas that more or less follows the third option described above (that evil has some divine purpose unknowable to humans), but this solution has not been highly regarded since the Enlightenment.

To answer the question about what concept of God will persuade future believers requires a look at some of the fundamental currents underlying contemporary thought. These might give us clues regarding what kind of solution may become plausible.

Just as a new idea often arises from the fading remnants of an old idea, so the rejection of the traditional notion of God by nineteenth and early twentieth century thinkers, such as Nietzsche, Feuerbach, Marx and Freud, was accompanied by a change in some fundamental assumptions about the nature of reality and the acceptance of new ideas. For them, the ideals of freedom and self-government obviated the need for the traditional "supreme ruler" God, who held the fate of His subjects within His power. Today, a new set of ideas is entering popular consciousness. These new ways of thinking suggest that a different notion of God may be in the process of construction in the contemporary Western world.

Two major ideas to imbue modern thinking come from science: the idea of evolution in biology and of relativity in physics. We can call these the Darwinian and Einsteinian insights, although they extend well beyond the speculations of these two men.

The fundamental idea behind evolution is that all things are constantly changing, that nothing stays the same. For some, this idea is a motive to reaffirm an image of God who is unchanging and eternal. But this would make God an exception to the metaphysical principles that govern all of reality. It would, in other words, place God further at the margins of understandability and availability. A God who is outside of time and change is a God who cannot intervene in history. Such a God would be quite useless, and believers in this God would be hard-pressed to draw a practical distinction between the consequences of their belief and that of atheists.

A better solution, one more in keeping with the Darwinian insight, would be to reconfigure the idea of God as temporal and changing. This would make God available for human concerns, one who can listen to prayers and respond appropriately. Curiously, such an image of God is not necessarily antithetical to divine omnipotence, at least according to one interpretation. To say that God is all powerful would not be taken in an absolute sense, but rather in a relative sense. God is as powerful today as He can possibly be, but He can (and does) exceed His own power at every point in the future. In other words, like us, God is changing, growing, evolving. He is always more than He was, and will always be more than He is. And at every moment, this changing God is omnipotent.

Note also the second, or Einsteinian, insight in this formulation. The divine existence is understood in a relative sense. God's power and God's love are relative to the requirements of the concurrent cosmic situation, and God, at any given moment in time, is limited to the exigencies of that situation. Divine immanence is stressed over divine transcendence. As the philosopher Alfred North Whitehead puts it, "It is as true to say that the World is immanent in God as that God is immanent in the World."[3]

Whitehead developed a notion of the "consequent nature" of God as encompassing all of reality, every puff of trivial existence, saved in a way to constitute God's concrete nature. A similar idea of God's relation to the world can be found in a grand synthesis developed by the French Jesuit thinker, Pierre Teilhard de Chardin, for whom God is all in all, the final cause of reality, overcoming all evil and drawing all things into his ultimate Self.[4]

This image of God is in some ways similar to the Eastern (especially Hindu) idea of "pantheism," which literally means that God is all, i.e., God is the totality of reality. Every bit of reality is a part of God; every event is a manifestation of divine Being. God is these things, not a cause of them and not separate and distinct from them. God is material because reality is material; God is in time because reality is in time.

The Western counterpart of pantheism, as expressed by Whitehead and Teilhard, for example, can better be called panentheism, which means that God is all, yet more than all. Like pantheism, it identifies

God with the totality of reality, but it also asserts that God is more than the sum total of everything. It is based upon the notion that the whole is actually more than the sum of its parts, just as a person is more than the sum of his cells or organs. In other words, the whole (God) is more than the sum of His parts (all the elements of reality), yet He is concretely constituted by these parts.

While few Americans would identify themselves as pantheist or panentheist if asked in a survey, I am convinced that this general way of thinking about God is becoming a point of convergence with implications for the future. I believe this is so because this idea of God touches so many important themes of contemporary thought.

Often one hears the word "force" in discussions about God. "May the force be with you," is the blessing given by Obi-Wan Kenobi to Luke Skywalker in the film Star Wars. For many, the problems inherent in the trilemma are not so persuasive as to require an abandonment of belief in God. But they are sufficiently problematic to cause a reconsideration of the traditional concept of God. For many, the result is an impersonal God, one who becomes the underlying force we experience as underlying all of reality.

This force is dynamic, changing. It is relative to, and perhaps one with, events as they emerge. Is this not what fervent Christians mean when they describe the work of the Holy Spirit?

Likewise, one often hears believers speak about the "presence of God." God is everywhere, in everything. No one can escape the presence of God. By this believers generally mean that God is present in all the area that surrounds things, but that He is a being distinct from these things and that His presence does not compromise the individual essences of things.

However, this belief in God's separateness is problematic, because these believers don't think of God as merely being in the spaces; they think of Him as immanent. "God is in you," they may say. But how far is this idea from the ideas of pantheism or panentheism: that God is all these things, or He is all these things and yet something more? In other words, the presence of God, a concept widely used by Jews, Christians and Muslims, can easily move towards an image of God in which He is identifiable with everything, the totality of reality, the universe in its ever-changing, relative state. God is everything, and everything is constantly changing. The presence of God is the force empowering that movement and the unity of the evolving universe itself.

Such a concept of God, as I have suggested, already has some resonance with believers, and may become more acceptable to them as they analyze practically what exactly happens when God interacts with them in their lives. Is it not through the events of time and nature that they believe God manifests his power (force?) to them? It also resonates with atheists and agnostics, who might loathe the word God, but who would be in accord that the universe is the sole field of action. So why all the fuss over a word?

Such a concept would surely find support among so-called "new-age" thinkers, who have borrowed heavily from the Hindu pantheistic philosophers in their world view of spirituality and interconnectedness. Finally, it would resonate most powerfully with environmentalists, atheist and theist alike, who would find that identifying God with nature provides the strongest possible basis for encouraging a profound respect for the nature of which we are a part.

To be sure, I have simplified this image of God egregiously. My purpose here is not to work out the philosophical and theological difficulties (and there are many!) inherent in the image I have suggested. Rather, I offer it merely as an insight, for whatever it's worth.

It seems to me that in our culture, it is harder to wrap our brain around an absolute God than a relative one; that God as a totally separate being is less appealing than an immanent one, and that an eternal God is not as religiously useful as a changing, evolving one. In other words, the absolute, transcendent, changeless image of God inherited from our ancestors may well be dead, or at least in its last throes. And we are loathe to embrace atheism. If this is so, then we need to reconceptualize God. The idea that our primary model for thinking about God should be the evolving universe itself makes a lot of sense.

ENDNOTES

1. From *The Complete Works of Friedrich Nietzsche,* Dr. Oscar Levy, editor (New York: The Macmillan Company, 1924), Vol X, pp. 167-169.

2. Shorto, Russell, "Belief by the Numbers," *The New York Times Magazine*, December 7, 1997, Section 6.

3. Whitehead, Alfred North, *Process and Reality* (New York: Macmillan Co, 1929), First Free Press Paperback edition, pg. 410.

4. Teilhard de Chardin, Pierre, *Le Milieu Divin* (London: Fontana Books, 1957)

DO UNIVERSAL HUMAN RIGHTS IMPLY THE FUTURE DEVELOPMENT OF A WORLD RELIGION?

by

Charlotte Waterlow

THE INTERNATIONAL LAWS OF UNIVERSAL HUMAN RIGHTS

After the nightmare of World War I, the League of Nations, that is, of sovereign nation states, was established in 1920 at Geneva to solve international disputes by arbitration. It included 31 states, but the USA refused to join it, and it was dissolved in 1946. Two subsidiary bodies remained in existence, the International Labour Organisation and the Permanent Court of International Justice at the Hague. This Court, however, had, and still has, no powers to enforce its decisions when governments submit disputes to it. Nineteen years later World War II, a far more terrible and comprehensive War, broke out, and when it was over in 1945 the *United* Nations was established. The word "United" has fraternal implications absent from the word "League". Its functions are set out in the preamble of its Charter: not only to stop wars but to develop "fundamental human rights" and to "promote social progress and better standards of life in larger freedom". The United Nations, when founded, had 58 member states. Partly as a result of decolonisation by the Western powers after the War, and the creation thereby of 80 new states, the United Nations now has 185 member states. Apart from one or two little islands, the only significant state which has refused to join is Switzerland.

After the Charter was signed in June 1945, an eight-member Committee was set up under the Chairmanship of Mrs Roosevelt—the only woman included—to draft the Universal Declaration of Human Rights, and it was adopted by the United Nations' General Assembly in 1947. Fifty eight member states voted 1400 times on practically every word and clause of the text. It is a statement of legal and ethical principles. I think that it can be regarded as the most important ethical statement since the Sermon on the Mount. Its terms were then embodied in two Covenants of Human Rights, one on Economic, Social and Cultural Rights, and one on Civil and Political Rights. These are treaties whose signatories commit themselves to implement the Rights. The second Covenant is accompanied by an "Optional Protocol" providing for individuals to appeal direct to a Human Rights Committee set up under its aus-

---

Charlotte Waterlow *is a retired high school teacher, Cirencester, UK.*

pices.   And there is another Optional Protocol advocating the
abolition of the death penalty.   The Preambles to both Covenants
"recognise that human rights derive from the inherent dignity of
human beings".

The two Covenants came into force as international law in 1976,
when 35 states had ratified them. The USA has signed and ratified
the Political and Civil Rights Covenant, but has not ratified the
Economic and Social Rights Covenant.   Saudi Arabia and South
Africa are almost the only other states which have not ratified the
Covenants.

Meanwhile the Council of Europe, a body of 40 European states,
now including Russia and ex-Soviet states, set up in 1949, drew up
two similar Conventions on Human Rights for Europe. The Council,
with its Parliamentary Assembly and Council of Ministers, is quite
distinct from and overshadowed by the European Community (now
called the European Union). But it has added to the Human Rights
system a unique and most significant feature: a European Court,
located in Strasbourg, to which citizens of the member states of the
Council which have signed the Conventions can appeal over the
heads of their governments, and the governments concerned are
committed to implement the Court's decisions. This is a beginning
of the development of the essential complement of law enforcement
procedures.   Another enforcement measure has been the
appointment, in 1997, of a Human Rights Commissioner. She is at
present Mary Robinson, the first President of Ireland, and she keeps
a special eye on the rights of women and the poor. Finally, an
enormous step forward has been taken in 1998 with the
establishment at the Hague of a World Court with powers to arrest
and try villains, in particular those responsible for genocide, over the
heads of their governments. The International Court set up in 1919,
whose constitution is incorporated in the United Nations' Charter,
can only deal with cases submitted to it by governments, and cannot
force the latter to comply with its rulings. The next step must be the
establishment of a world police force - NATO provides the outline
of a model.

THE ETHICAL BASIS OF UNIVERSAL HUMAN RIGHTS

"Human Rights" is therefore a new concept in the annals of world
history. It affirms the *right* of every *individual person,* of whatever
sex, colour, social status, and intellectual ability, to be endowed by
society—which today means generally the nation state to which he
or she belongs—with the conditions to enable him or her to express
his or her unique qualities as a person. Article 26 of the Universal
Declaration hints at the spiritual implications of this philosophy. It
states that "everyone has the right to education . . . . . Education
shall be directed to the full development of the human personality".
What proportion of the human race today has the opportunity fully

to develop his or her creative potential whether he or she lives two years or 80 years? I will return to this point.

Human Rights are *ethical* standards, concerned with right and wrong. Are ethics subjective, attitudes derived solely from the human psyche, or are they objective, attitudes derived from the person's perception of ethical standards and principles which stem from a divine source—God? I recently encountered a former high school pupil, the son of a Harvard Law School Professor, who was taking a post-graduate course in ethics at the Harvard Law School. I asked him what he thought about this matter. "My professor thinks that ethics are objective, but I think they are subjective", he said. He is not alone; there is much confusion about the nature of ethics in the modern world. First, the great traditional religions have different ethical systems (see below). Secondly, in the modern age, for the first time in history, atheism and agnosticism, denying that ethics stem from a divine source, are widespread. In the pre-modern world, that is, essentially before the 18th century, only a few eccentric Greek philosophers were agnostics (a Greek word).

In 1893 an unprecedented event occurred: a "Parliament of the World's Religions", composed of religious leaders from all over the world, met in Chicago. In particular the first leading Hindu sage to come to the West, Vivekananda, made a profound impression; his coming heralded a phenomenon of immense importance; the inflow of Oriental religion into the Western psyche. In 1993 a centenary meeting was held, attended by nearly 7000 religious people—a further 3000 had to be turned away. The essence of the Parliament's conclusions is summed up in a booklet edited by the enlightened German theologian, Hans Küng, called "A Global Ethic".[1] The Parliament affirmed that there can be "no global order without a global ethic", and that this ethic is objective, in the sense mentioned above. "As religious and spiritual persons we base our lives on an Ultimate Reality" (pages 18-19). The global ethic should be based on four irrevocable directives: 1. Commitment to a culture of non-violence and respect for life. This includes disarmament and respect for animals and plants. 2. Commitment to a culture of solidarity and a just economic order. 3. Commitment to a culture of tolerance and life of truthfulness. And 4. Commitment to a culture of equal rights between men and women. These commitments involve a "transformation of consciousness, through reflection, prayer, meditation or positive thinking" for "a conversion of the heart". The Buddhists suggested the titles "Great Being", "power of the transcendent", or "Higher Spiritual Authority" instead of "God" in reference to "ultimate spiritual reality". (See the titles for God used by Thomas Jefferson in the USA's Declaration of Independence, quoted below). This moving pamphlet was endorsed by the representatives of all the great religions at Chicago.

Nevertheless, the global ethic of the Parliament of Religions does not answer the question which I put to my former student: what is the source of ethics? I shall suggest below that the answer to this question is crucial to the implementation of Human Rights, and that this implementation is crucial to the future of the world. When the Universal Declaration of Human Rights had been drawn up the Soviet delegate said to Mrs Roosevelt: "If the Declaration mentions God, the USSR cannot sign it". But the Pakistani delegate said to her: "If the Declaration does *not* mention God, Pakistan cannot sign it". In fact it does *not* mention God and Pakistan did sign it!

TRADITIONAL RELIGIONS IN THE MODERN WORLD

In spite of this unprecedented and magnificent structure of international human rights law, based on ethical principles; in spite of the growing movement for the great religions to confer together and to draw up a universal ethical code (see "Faith and Interfaith in a Global Age" by Marcus Braybrooke, Joint President of the World Congress of Faiths[2], the past fifty years has been a period of international, national and personal violence and social and economic injustice on a monumental scale. The reason may be that religion has been left out of the emerging global order - and necessarily so. For some 5000 years several major religions dominated the political, economic and social scene of great Empires, laying down what all classes of society were to believe, and how they were to behave, controlling the social structures, which in most cases remained unchanged for hundreds or thousands of years—until the end of the 18th century, when the great clarion call of the French Revolution, "Liberty! Equality! Fraternity!" rang out, calling for fundamental change. In order to understand what is happening today and could happen tomorrow we must glance at these ancient structures.

Moving from east to west, major civilisations arose in Japan, China, Mesopotamia, Palestine, Egypt, Greece, Rome, Western Europe after the collapse of the Roman Empire, Byzantium, Russia, the Muslim Empire founded by the successors of the camelherdsmen, and Central and South America. All except those of Greece and Rome (which were only partial exceptions) were dominated and moulded by a public religion to which all citizens were expected to adhere. The rulers were often regarded as gods, or as God's representative on earth. The coronation ceremony of the present Queen Elizabeth of Great Britain, held in 1953 in Westminster Abbey was a relic of this attitude. After several hours of ceremony she was invested with the crown, symbol of spiritual authority, first worn by King Edward the Confessor, who ruled from 1042-1066. The Archbishop of Canterbury then anointed her as she held the orb and sceptre. The present Emperor of Japan, whose ancestors were regarded as divine until 1945, was consecrated in 1990 in a four day ceremony of probably prehistoric origin. It culminated in two ritual meals where the Emperor sat facing two empty seats, at which the *Kami*, ancestral gods, were supposed to be seated. Scholars agree that "the ritual sequence enables the Emperor to pass from a human to a divine condition".[3]. . . . Compare these ancient ceremonies with the inauguration of President Clinton, in 1993 dressed in an ordinary dark business suit similar to that worn by the majority of his countrymen, standing in the open in front of the Capitol (the US Legislature), and in one short sentence swearing to do his best to "preserve, protect and defend the Constitution of the United States". In traditional societies religious laws and customs, believed to stem from a divine source and in most cases laid down in scriptures (those of Judaism, Christianity, Buddhism, Hinduism and Islam are the main examples) prescribed what each class of people should do and how they should behave. The Hindu caste system, the Judaic rules about food, the Muslim rules about womens' clothing, the Christian and Muslim prohibitions of usury, the circumcision of African girls, are a few examples from a collection of thousands.

> The rich man in his castle,
> The poor man at his gate,
> God made them high or lowly,
> And ordered their estate".

Except in ancient Greece and Rome, which had scripts but no scriptures, and whose leaders were preoccupied with philosophy rather than religion, this static attitude laid down by the scriptures prevailed throughout the world for millennia. It is summed up by Shakespeare in a famous passage in "Troilus and Cressida" about "degree", or each thing in its appropriate place.

> "O, when degree is shaked,
> Which is the ladder to all high designs,
> The enterprise is sick! How could communities,
> Degree in schools and brotherhoods in cities,
> Peaceful commerce from dividable shores,
> The primogenative and due of birth,
> Prerogative of age, crowns, sceptres, laurels,
> But by degree, stand in lawful place?
> Take but degree away, untune that string,
> And hark, what discord follows!. . . "

Normally 80% of the population were peasants, tilling the land or fighting for nobles and king, or building edifices from the Great Wall of China to the medieval churches which spatter Europe - and a myriad other monuments all over the world. Their status was that of slaves, serfs, sharecroppers, and sometimes small landowners and craftsmen. Their expectation of life was about 30 years. In the 16th century the English philosopher Thomas Hobbes described their lives as "nasty, brutish and short". Above them were the nobles, who owned and administered the land for the ruler. Sometimes, like the

Brahims of India and the Mandarins of China, they performed priestly functions, but above all they fought for him. Kings and nobles as well as peasants were often illiterate. Women were housewives and field workers. During these centuries there was very little scientific or technological development, except in mathematics. Printing was invented in China in the sixth century A.D. and in Germany in the 15th century! Cicero (B.C. 106-43) and Lord Byron 2000 years later travelled around in similar horse drawn carriages with beds in them. Horse back was still the fastest mode of travel. Giordorno Bruno was burnt at the stake in Rome in 1600 A.D. for teaching about the infinity of the universe and the plurality of souls, just before Galileo acquired his telescope and opened up the spectacle of the countless stars in the Milky Way. Since the Bible is silent on astronomical matters, the Roman Catholic Church had adopted the Greek teaching that the sun goes round the earth. Thus for several thousand years it was religion - the Will of God or the gods, laid down in the infallible, canonical scriptures, interpreted by the priests who were also the jurists, which kept everyone in his lawful place or "degree". The penalties for flouting the rules of your degree, in particular for questioning the scriptures, were dire. Mary Tudor, the Roman Catholic Queen of England from 1553 to 1585, burnt Protestant bishops at the stake, including Archbishop Cranmer who produced the first English Protestant prayer book. In 1922 A.D. Mansur al Hallal, an Iraqi wool carder of the Sufi mystical sect of Islam, was crucified for heresy because he claimed "to be God". He meant that he felt himself infused with the light of God (see IV below). When he saw the cross and nails he prayed aloud to God for forgiveness of his executioners, saying: "For verily if Thou hadst revealed to them what Thou hast revealed to me, they would not have done what they have done; and if Thou hadst hidden from me that which Thou hadst hidden from them, I should not have suffered this tribulation".[4] In the case of Islam the Holy Koran, regarded by all devout Moslems as infallible, extends these penalties to the next life. There "the faithful and the unbelievers contend about their Lord. Garments of fire have been prepared for the unbelievers. Scalding water shall be poured upon their heads, melting their skins and that which is in their bellies. They shall be lashed with rods of iron. Whenever, in their anguish, they try to escape from Hell, the angels will drag them back, saying "taste the torment of Hellfire!"[5]

Even today, in Saudi Arabia, where Human Rights have not been enacted, women are stoned to death by law for committing adultery. In the 18th century the "Age of Enlightenment" began to dawn in Western Europe. The famous French philosopher Voltaire cried out, "Écrasez l'infâme!"—"Crush the vile thing!"— the Catholic Church. Rousseau wrote his famous book "The Social Contract" about the rights of "the people" to overthrow a bad government; and advocated deism (worship of God but not of Jesus Christ), republicanism and the rights of women. Thomas Jefferson, influenced by Paine, drafted the United States' Declaration of Independence, which asserted that "the laws of Nature and Nature's God" give all people, as equals, "certain inalienable rights", including "Life, Liberty and the pursuit of Happiness", with which "they are endowed by their Creator". Washington, Jefferson and other leading Americans and Europeans became Freemasons, rejecting Christianity altogether. The Freemasonic symbol still adorns the dollar bill, and one of the most marvellous musical works, Mozart's opera "The Magic Flute", is based on Freemasonic mystical ritual.

And so the modern age dawned, the age of science and Human Rights. In two short centuries these completely new developments have transformed Europe and North America, and partially or largely transformed the rest of the world.

The Age of Enlightenment also involved the rapid, unprecedented development of science, both pure and applied, ushering in for the first time in history the concept of a *dynamic* society, based on such concepts as "development", "progress", and "growth", which were the product of democracy and personal freedom. The modern age had dawned. Meanwhile in the atmosphere of intellectual freedom necessary for science, agnosticism and atheism became general in the Western world. In the 20th century the secular Enlightenment initiated in the West has been enormously enhanced by the flood of new light from the East. The mystical scriptures of Hinduism and Buddhism were translated by Western scholars under the auspices of colonial rule and flooded into Europe and North America, infusing or confronting the congealed dogmatism of Christianity. Spiritual experience was felt by many to be more important than intellectual knowledge of the scriptures and rituals, which seemed irrelevant to modernity. The nut of Islamic law and practice is proving harder to crack.

As he watched the French Revolution develop into "the Terror" the poet Wordsworth groaned. He saw the concept of what Shakespeare called "degree" being smashed up. He wrote:

> "But a terrific reservoir of guilt
> And ignorance filled up from age to age,
> That could no longer hold its loathsome charge,
> But burst and spread its deluge through the land".

Shelley also had a premonition of the darkness which must accompany the light of the Enlightenment; his short poem "Hellas" is a *cri de coeur*. For in spite of the creation of the magnificent structure of international Human Rights law, outlined above, in spite of the movement of the great religions to confer together, and draw up a universal ethical code, the 20th century has been a period of international and national violence on a monumental scale, and religion has been left out of the emerging global order—and necessarily so. For this new order stirs up immense ignorance and fear in these ancient systems.

"Fixed opinions are like standing water, they breed reptiles of the mind", wrote the poet William Blake (1757 - 1827). The reptiles are called "fundamentalism", the urge to challenge the ideas and practices of modernity by proclaiming, often violently, ancient dogmatic teachings. Imagine a simple Iranian cleric or imam in a remote village or small town. He speaks the national language, Farsi, and he reads script, the script of the Holy, infallible Koran and of Farsi. He picks up bits of national and international information from the village radio or T.V. But the great modern Western secular "subjects" such as economics, sociology, psychology and science in its many aspects, are a closed book to him. The USA is called "The Great Satan" because it represents the Western consciousness, expressed in the collectivity of these subjects. It cannot be reconciled with the belief that all truth is in the Holy Koran. An Egyptian woman scholar whom I met in the USA told me that very little Western literature in these fields has been translated into Arabic, and the translations are bad. Most Egyptians who want to study these subjects read about them in English or French.

Fundamentalism in the USA seems to be essentially a revolt by the poor and despised, and particularly the Blacks, against the rich and powerful Whites. It will be recalled that the USA has not yet signed the United Nations' Covenant of Economic, Social and Cultural Rights. The Christians who dominate the political and economic life of the country have only implemented what in West Europe is called "The Welfare State" to a very limited degree. Many Blacks, in their frustration, have turned to Islam. In the 1920s they formed "The Nation of Islam", led by a Black peddler called Wallace D. Fard, who was declared by his followers to be Allah, or God. He was succeeded by Malcolm X (1925-65), who rejected Martin Luther King's non-violent approach.[6]

American and European fundamentalist movements are sad but minor phenomena. But fundamentalism on a major scale may threaten the very future of humankind. There are a billion Muslims in the world, and the great majority live in the region which stretches from Indonesia and Afghanistan to Morocco. Many are in the mental state of an Iranian or Pakistani village imam described above. They are filled with resentment towards "The Great Satan", but they possess 80% of the oil which the industrialised countries need, and in return they import from these countries "weapons of mass destruction". In 1997 the United States supplied 42% of the global arms market—three quarters going to the Middle East.[7] The spread of religious fundamentalism, particularly in Muslim countries, could destroy the world.

4. THE WORLD RELIGION OF THE FUTURE

This brings us back to the question of whether ethics are subjective or objective. If they are regarded as subjective, Human Rights may not be taken very seriously by those whose interests they challenge, whether politicians who want to make war or business men and women who want to make money. The failure to halt the spread of "weapons of mass destruction" or to alleviate the huge gap between the 80% of the world's population whose "basic needs" are not fulfilled and the 20% who have more than they need shows that global adherence to the concept of Human Rights is largely intellectual. Despite the exploding number of world Charities (Non-Governmental Organisations) hearts have not been touched on a wide scale. Religion therefore remains traditional and dogmatic and fails to be the engine for the implementation of Human Rights.

If ethics are objective there are two alternatives. The first is to try to relate them to the traditional religions, which may involve doctrinal clashes, or even the rejection of Human Rights Covenants and Conventions, as in the case of Saudi Arabia and the USA. The way forward for the future lies in the development of what we might call the cosmic consciousness, or mysticism—the full powers of personality, hinted at in Article 26 of the Universal Declaration of Human Rights. I have already described how this state of consciousness was achieved by a medieval Iraqi wool-carder - only, alas, to lead to his being burnt alive. A similar experience, but with a happy ending, was attained by one of the greatest European Christian theologians, St. Thomas Aquinas. After spending many years studying and teaching, with the aim of integrating Aristotelian Greek philosophy into Christian theology, he one day had a mystical experience while attending Mass. After this experience he put down his pen, saying: "Such things have been revealed to me that now all I have written appears in my eyes as of no greater value than straw". And he never wrote another word! Mysticism involves filling the individual consciousness with the light and love of the Creator (assuming that He/She objectively exists!). And in institutions it leads from law to "community" or "fraternity". In the future the concept of Human Rights will either tend to remain a collection of intellectual concepts and laws, or it will prove to be the ethical engine inspiring the development of a world Community based on personal experience of the Light and Love of the Creator, in whatever theological or philosophical terms.

It is significant that Kofi Annan, the present Secretary-General of the United Nations, has re-established the Meditation Room in the United Nations building, which was first set up by Dag Hammorskjöld, its second Secretary-General who died in an air crash in 1961. The room contains only a large Scandinavian stone in the middle, upon which shines a shaft of light from above.

REFERENCES

1. Hans Küng and Karl Josef Kuschel, *A Global Ethic: The Declaration*

of the Parliament of the World's Religions. London: SCM Bros. Ltd, 1993.

2. Marcus Braybrooke. *Faith and Interfaith in a Global Age.* Oxford, England: Braybrooke Press, 1998.

3. Carmen Blacker. "The Shinza or God-Seat on the Daijusain Throne." *Bed or Incubation Couch (Japanese Journal of Religious Studies)*, 1990, p. 179-97.

4. Charlotte Waterlow. *The Hinge of History.* London: One World Trust, 1995, p. 35.

5. *The Koran.* Penguin Books, England.

6. Ninian Smart. *The World's Religions.* Cambridge University Press, p. 374.

7. Figures from Campaign Against the Arms Trade, London.

GROW OLD ALONG WITH ME

by

David Macarov

A BRIEF REVIEW CONCERNING PREDICTIONS

People have been trying to foretell the future from the earliest days of recorded history. In some places and at some times there were medicine men and shamans, witch doctors and seers. They were supplemented by inanimate objects, of which the Greek oracle at Delphi is perhaps the best known. And then there was Cassandra, whose name has become a by-word for fearsome prophecies, who always predicted correctly—and who always saw her predictions ignored. Many of the Hebrew prophets discussed events to come, some at the end of time, but most for shorter periods. The Middle Ages saw Nostradamus reading the stars and making predictions from them—a tradition that continues both in modern astrology and in continuing quotations from that medieval astrologer. Today, the advent of every new year is marked by published predictions by better—and lesser-known seers, whose occasional coincidental triumphs are trumpeted, while their many failures are ignored.

During the last fifty years, however, there has grown up a school of thought, if not a discipline, which makes use of more rational, exact and replicable measures of the future. These measures include the consensus of experts, mathematical equations, computer simulations, content analysis, analogies and comparisons, trend predictions, and scenarios. Most current projections into the future make use of combinations of these predictive methods.

What has been the track record of such future predictions? Not bad. Of the 134 inventions or changes envisioned by George Orwell in 1946 as occurring by 1984, over a hundred did indeed come to pass, and more have been added in the meantime. In an article in the 1997 issue of The Futurist , it was found that 68% of the predictions made thirty years ago had come true.[1]

Clearly, some expected changes are more important than others, just as some are more probable than others. It is therefore possible to weigh the possibility of change against the importance of the change, in somewhat the following manner:

Possibility of Change

- Impossible
- Barely possible

David Macarov is emeritus professor at the Paul Baerwald School of Social Work at Hebrew University, Jerusalem, Israel.

- Possible, but not probable
- Probable
- Very probable
- Almost certain

Importance of Change

- Catastrophic/Utopian
- Extremely important
- Important
- Not very important
- Unimportant
- Trivial

A result that will be extremely important deserves attention and action even if the possibility of its occurrence is not highly probable; conversely, a result that is almost certain to occur, but whose consequences will be trivial, does not call for concentrated planning or action.

Predicting the future would be merely an academic game if such predictions did not generate attempts to influence the change itself, or to deal with its consequences. As Herodotus said, "The bitterest of all griefs—to see clearly and yet be unable to do anything." In order not to be in that position, the 68th International Conference on Social Welfare, held in Jerusalem in 1998, invited representatives of four non-governmental worldwide organizations dedicated to change to present their methods of operation and their results. These included UNICEF, which works through publicity and pressure; Greenpeace, which is an activist organization; the World Health Organization, which teaches, measures, and engages in start-up operations; and the International Monetary Fund, which uses financial pressure. There are many other organizations trying to influence the future, such as those dedicated to saving endangered species and those opposed to child labor and to slavery, that have had at least partial success in achieving change and thus influencing the future.

As indicated above, some changes are more certain than others, and some are more important. One of the most certain changes during the next fifty years at least, and one with enormous ramifications throughout the social, economic, ecological and governmental spheres, is the continuing growth in the older population. In the chart shown above, the relationship would be a diagonal line from the bottom of the left column to near the top of the right.

Given the ramifications of a relatively large increase in the number of the elderly, there are many efforts being made to anticipate the situation. Although it does not seem possible to limit or influence the trend, there are some indirect forces at work. At present, fertility rates are falling throughout the world, and to the extent that this continues, there will eventually be a contraction of populations generally, and thus a lessening of the number of the aged, although the proportion may remain the same or even grow, given continuing lengthening of life expectancy.

Insofar as dealing with the effects of the burgeoning aging population is concerned, many organizations and bodies are active. For example, the United Nations has pronounced 1999 as the Year of the Aged, and there are numerous conferences, research projects, and programs dealing with various aspects of improving the forthcoming position of the elderly.

WHO IS AGED?

The age at which someone becomes elderly can only be drawn arbitrarily, and for most practical purposes this is considered to be the age at which pensions begin to be paid. When Chancellor Bismarck established one of the first social security programs in Germany in the 1880s, he set age sixty-five as the pensionable age. Since not many people lived to that age at that time, he was not taking a large economic risk. In any case, sixty-five became and has remained the most popular age for the beginning of old age pensions in the developed countries, even though life expectancy has lengthened considerably in the meantime.

But these programs account for only 28% of the old-age pension programs in the world. In developing countries, whose programs constitute 71% of the world's total, life expectancy is much shorter. In many African countries life expectancy may be as short as age 33, as in Sierra Leone,[2] or 41, in Guinea-Bassou, or 43, as in Burundi, Gambia and Swaziland. Consequently, 60 is the most common pensionable age in developing countries. However, 30 countries, or 19% of the world total, pay pensions from age 55. There are also some countries that begin paying pensions at age 50, although this is often limited to the "prematurely aged," or people in hazardous occupations. A few programs pay pensions from age 45, contingent upon actual retirement from employment or following a sustained period of unemployment. The only country in which retirement pay without conditions begins at age 40 is the small Pacific island of Kiribati. There are some interesting exceptions to the age factor. In Sri Lanka a female employee who gets married is entitled to a pension, and in the Czech Republic retirement age is lowered for women, depending on how many children they have. Generally speaking, however, there seems to be a linkage between life expectancy and retirement age.[3]

Despite the fact that full pensions are paid to males at age 65 in the United States, over 90% of all male retirees there are currently giving up three years salary and 20% of their pensions for the rest of their lives by taking early retirement at age 62; and almost three-fourths of all persons receiving Old Age, Survivors and Disability Insurance (OASDI) retired early.[4]

Thus, there is a difference between the age at which (full) pensions are paid, and the age of actual retirement from work. Further, some organizations define the aged differently. The American Association of Retired Persons accepts members from age 55, as do Elderhostel study courses. These facts raise some obvious questions about counting the aged in society as only those 65 and older. If old age were to be considered to begin earlier than age 65, then the number of the elderly in society today, and to be expected in the future, must be greatly enlarged, accentuating many future changes in this area.

HOW MANY AGED?

Regardless of definition, the future concerning the elderly is not just a guess. Those who are already old and those who will become old are now alive, and their life expectancies can be predicted with a reasonable degree of certainty. As life expectancy continues to extend due to improvements in sanitation, diet, medicine and housing, among other things, so will the number of the aged. In the sixteenth century B. C., life expectancy in Egypt was only 15 years.[5] The Romans lived for about 30 years[6] and in 1700, life expectancy in Europe was still only 33 years. At the beginning of the present century it was about 40 years, and in 1950, it was 66 years. In 1990, it was over 74 years.[7] It has been estimated that by the year 2050 there should be no country with a life expectation of less than 50, and that the current difference of 42 years between countries with the highest life expectancy (Japan) and those with the lowest life expectancy (Sierra Leone) will be reduced to 31 years.[8] There is also the prediction that the human life span will increase to between 120 and 130, and possibly longer.[9]

It should be noted that in the developed countries women tend to outlive men by about seven and a half years, while in the developing countries, this gap is only two years, and sometimes disappears altogether. In the future, increasing longevity will take place at a higher rate for women than for men, and the present gap can be expected to widen, at least until the year 2020.[10] A number of factors contribute to the fact that women tend to outlive men, but a United Nations publication says: "Cigarette smoking is probably the principal factor affecting the sex differentials in life expectancy."[11]

Parenthetically, it should be noted that although infant mortality rates affect the statistics of life expectancy, current predictions are borne out by studies of adults at various ages.[12] An American of 80 can expect—on the average—to live another five to seven years.

In nearly all European countries (Ireland is one exception) the aged population in 2020 will be more than 50% larger than it was in 1960.[13] In the United States the proportion of those over age 65 in 1900 was 4%; in 1984 it was 11%; and it is estimated that by year 2020 the proportion will be over 21%. That is, in 1900 one person in 16 in the United States was over age 65; in year 2030 it will be one in four.[14] Finally, in terms of numbers, the largest growing group among the aged, proportionately, is the 85 plus group. This group numbered two and a quarter million in the United States in 1980. It totals about five million now, and will grow to 13 million by the year 2040.[15]

Some probabilities

The growth to be expected in both the number and the proportion of the aged in future society will have very important implications for many societal institutions and for the individuals concerned. A partial list of these would include the areas of health, economics, and family.

HEALTH

Most of the elderly will be much healthier than their predecessors were, and for a much longer time. As medical science continues to progress, and as knowledge and practice concerning healthier diets and methods of warding off heart attacks and strokes spread, many of the ills afflicting the current elderly will be obviated or easily cured. For example, cataract operations which were once almost major surgery, followed by a long period in which the head could not be moved, have become in-office procedures using non-invasive local anesthesia, laser beams and next-day recovery. Equal advances concerning other conditions will undoubtedly follow, probably most importantly in the area of genetic manipulation.

These many additional years of non-working time for the healthy majority of the aged will undoubtedly result in an enormous increase in recreational activities. Golf and tennis facilities will expand in countries where these are favorite sports for older persons. Bowling, lawn bowling, folk- and ballroom dancing and many other more-or-less sedentary activities will grow in popularity. Spectator sports will also draw an increasingly large group of older fans. Television viewing will occupy greater portions of the older persons' time, and movies may make at least a limited comeback. The use of the internet for recreational purposes is no less widespread among the elderly than in the rest of the population. There has already been a steady increase in travel by the aged, including cruises, and many recreational firms and organizations are targeting the elderly as their best potential source of clients. There has also been a growth in educational opportunities for senior citizens, including the growth of Elderhostel programs, and the same is true of various kinds of service programs. Indeed, the great bulk of volunteers—especially in social service organizations—will be aged persons. With the exception of personal and family problems, Woopies (Well Off Older Persons) may need little outside help.

Most of the recreational and leisure-time needs of the elderly will continue to be provided by private market sources. The need for tour managers, hotel staff, administrators of country-clubs, and like personnel will grow. Although well-trained and organized, recreational and leisure-time professionals will be greatly outnumbered by untrained or semi-trained personnel. The market for prosthetic devices, helpful household gadgets, electric carts and other such help to activities of daily living will increase. However, to the extent that sufficient medical progress is not made for those who suffer from long-term debilitating diseases, such as Alzheimer's, Parkinson's, and Multiple Sclerosis, there will be need for care, mostly in the home, over much longer periods than in the past—twenty, thirty and even forty years. The traditional source of such care is the eldest daughter,[16] but two recent changes operate to make this more difficult than it was in the past. The first is the number of women entering the work force, and thus unable to stay home with an elderly parent, and the second is that as life continues to lengthen, the daughter herself may become old enough to need help. There are increasing cases of the elderly taking care of the very old—the 70 year old daughter taking care of the 90 year old mother, for example. At these ages both of them might need care with their own activities of daily living. In addition, there are the psychological stresses and strains that often accompany family members taking care of their own relatives, raising the question of who cares for the carers? In these cases, professional help becomes indicated, and, indeed, the number of social workers dealing with the aged has been increasing, and will probably continue to do so.

One relatively recent method of trying to deal with this problem of shortage of carers has been the growth of sheltered housing for the aged. People thus move into private rooms or apartments in sheltered housing while still comparatively young and healthy, and are offered the physical, social and medical support they need as they age. Unfortunately, such housing is usually quite expensive, and although in some cases it serves additionally as an investment in real estate, in most cases the inhabitant loses money over time, and often loses all the money invested.

There is a new trend taking place, however, and this is the use of paid carers who live with the older person in his or her own home, thus obviating the need to change living quarters, amenities, neighborhoods, etc. So strong is the desire to remain in one's own home, even with advancing age, that the modal age at which a widow moves into some sort of sheltered housing is seventy-nine.

Many such carers are from the relatively unskilled strata of society, and as the job market becomes more and more bifurcated—a minority of good well-paying jobs at the top and a majority of low-level low-paying jobs at the bottom—employment opportunities in this area might grow considerably.

In some countries the need for caretakers for the elderly is met by importing helpers from the developing—or, at least, less developed—countries. In Israel, for example, carers—mostly from the Philippines—are brought in legally, paid a minimum wage, and have the protection of the social welfare and legal systems. One result of the growth of this movement has been a decrease in demand for sheltered housing, and a consequent decrease in building it. Indeed, in many cases sheltered housing is now being offered on more and more attractive terms to meet the competition from foreign carers.

As medical technology offers more help to the ill and longer life to the healthy, health-care costs increase. Medical costs for the aged average six times more than that for young adults, and this is expected to increase.[18] It is not uncommon for people who have lived reasonably prosperous lives to have all of their assets and even their incomes wiped out by a long-term illness, or a complicated operation, or even, in some cases, by the cost of diagnoses through a series of Catscans, MRIs, and other expensive procedures. This is despite the growth of medical insurance programs like Medicare and Medicaid.

In summary, there will be continuing growth of the number and proportion of the aged in the population, regardless of how they are defined, and the number of the very old will grow disproportionately. The great bulk of the aged will be able to find their own way to the activities in which they want to engage, subject to economic limitations, but those who develop disabilities associated with age will need help for much longer periods than at present. Provision of such help will take many forms, but may provide employment for otherwise low-skilled people, and for migrant workers imported for this purpose.

MENTAL HEALTH

There is a widespread mythology that once people retire from work they become depressed, unhappy, antisocial, and physically ill. However, there is a great deal of evidence to show that happiness after retirement is dependent, in large part, on sufficient income to live without undue stress. For those without economic strain, life after retirement consists of what Laslett calls "agentic self-fulfillment" —pursuing one's own projects and planning one's own life.[19]

The fact, noted above, that over three-quarters of male retirees in the United States are retiring early hardly supports the feeling that people resist retirement, especially now that dismissing an employee because of age has become illegal. There has been no diminution as a result of such laws in the number of people seeking early retirement.

Retirees who have enough money on which to live decently generally report themselves as happy in their retirement, glad they retired, and sorry they had not retired earlier.[20] Men who retired close to their expected retirement age did not differ significantly in their reported happiness from those who continued working. Indeed, even without regard to retirement, most people over 65 say they are satisfied with their lives. One-third of them call the later years the best years of their lives; only a quarter report their situations as "dreary;" and just 13% say they are lonely.[21]

Nor does retirement necessarily lead to physical deterioration. Although 10% of blue-collar retirees polled indicted that their health declined after retirement, twenty-five percent reported actual improvement. In one study, all retirees who had suffered from ill health reported an improvement upon retiring.[22]

Part of the anti-retirement mythology holds that the lack of fixed time schedule—e. g., preparing for work, going to work, certain days for work and others as free—is one of the most difficult readjustments among the retired population. However, as Kaplan[23] points out, people experiment with leisure time activities during weekends, holidays, vacations, and after working hours before retirement, which eases them into full leisure time with little difficulty. Experience also indicates that time schedules, although different, are re-established very quickly, including recreational activities, favorite television programs, using the internet, family visiting, naps, etc.

There has, in fact, been a burgeoning of new activities as older people begin to draw, to paint, to write and to contribute their time to various volunteer projects. One can only surmise as to how much more creativity would have been released, and how many more contributions to society might have been made, if retirement had taken place earlier; or, if the financial situation of all the poor had been such that they could participate in such activities.

ECONOMICS

For the non-wealthy among the elderly problems of income maintenance may become more intense. As it is, the largest single group among the poor in most countries is the elderly population.[24] There are differences in the statistics, since governments use various methods of defining poverty. Generally speaking, poverty can be defined in absolute, relative, normative, and subjective terms, and can be based on either income, assets, or both.[25] The elderly constitute 57% of the poor in England, 40% in Australia and 36% in Israel, according to governmental definitions.[26] However, if 60% of the average wage is used as the poverty line, then the number of the aged in poverty ranges from 60% in the Netherlands and 15% in Sweden to 34% in the United States and 58% in Australia.[27] As governments struggle with the problem of how to maintain promised pension payments in the face of growing beneficiaries and declining contributors, the very real fear of decreased pension payments causes anxiety among the elderly. These fears are not unfounded. It is projected that by the year 2000, American Social Security will pay only 54% of the pre-retirement year's wage to low earners; 42% to average earners; and only 28% to maximum earners.[28]   This is compounded by the fact that 40% of the aged in the United States has no prospect of a pension other than through Social Security; one in five household has no fixed assets; and one in seven has no health insurance.[29]

The large drop in income on the day that a male becomes 65 years old is sometimes rationalized on the basis that older people have fewer expenses, and already own the assets that they will need. Leaving aside the enormous medical costs with which many of the elderly are, or will be, burdened—as noted above—the fact is that the costs on day 65 plus one are not much less than they were on day 65 minus one, which is why most people living only on social security experience real drops in their level of living.

Insofar as living on assets is concerned, those who own their homes could probably supplement their incomes by selling their property and converting these assets into an income stream, but in very few cases would this be enough to pay for long-term residential care for any length of time.[30]

The problem facing all retirement programs, governmental, fraternal or private, is how to maintain payments to a growing groups of recipients for a longer than anticipated period of time. This will probably become more difficult, since no life insurance or pension program in the world was or is actuarially based on the anticipation of people living to age eighty or more.

The simplest solution for the economic problem of retirement programs would be to raise the retirement age considerably, or to eliminate it entirely. However, this proposal runs head-on into one of the major and most intractable problems facing modern governments—that of unemployment. If older persons remain at work, job opportunities for younger people are proportionately limited. It is for this reason that the United States government—seized of the problem of the liquidity of Social Security in the future for some time—has responded by gradually lengthening the age at which pensions will be paid by only two years, and that long in the future.

As Gal points out, changes in social security systems will probably take the form of new non-means tested and non-contributory benefits for new groups of beneficiaries, the introduction of more stringent eligibility conditions for existing programs and a greater dependence upon private markets.[31] Means-testing may be introduced for the elderly, especially at the higher income levels.

It is likely that pension payments will be reduced, as that is the easiest method of dealing with the problem, despite its political ramifications. Such reductions may not come under the guise of cuts in payments, but in reductions in cost-of-living allowances, payments made later, and more stringent requirements for payment. As it is today, almost 90% of social security programs for the elderly throughout the world are only for workers.[32] This means, in most cases, that women who have not worked, entrepreneurs, those incapable of working, and those who could not find work or could only find part-time low-paying intermittent jobs, are omitted from the pensionable group.

On the other hand, concomitant with the growth of the elderly in society will be their political and purchasing power. Few countries have, as yet, political parties based upon age, but there are many political lobbies and pressure groups representing the elderly. The largest volunteer organization in the world today is the American Association of Retired Persons, which also includes members from other countries. Although eschewing a political role in terms of endorsing candidates or parties, it nevertheless takes a very active political role in lobbying for the interests of the aging and thus, indirectly and tacitly, for candidates who support such measures. A severe turn for the worse in the condition of the aged could see such an organization becoming a political power.

Similarly, growth in the purchasing power of the elderly will force changes in products, merchandising methods and even in the fiscal system. Products will continue to be geared to the elderly more and more, from patented bottle openers through personal alarm systems to housing built with age in mind. Advertising will use older models, larger print, and less sexy symbols. Banking, investment and real estate firms will be dealing with a clientele not investing for their old age or for a rainy day.

In summary, poverty among the oldest stratum of society will continue to grow as social security programs of various kinds take steps to defend their liquidity and as medical costs grow at an unprecedented rate, but although many aged people will be poor, most old people will be middle-class or better, and altogether they will be a political and economic power to be reckoned with.

FAMILIES

At one time it was easy to define a family: A working father with a wife and children, living together, constituted a family. One of the most telling statistics concerning recent social change has to do with the fact that this now describes less than 6% of American households.[33] The variety of households nowadays is legion. For example, there are extended families living together, multigenerational families, single-parent families (including both widows, widowers, and divorced persons, as well as single unmarried parents), unmarried cohabiting partners with and without children, childless families, families with one or more remarried spouses, same-sex couples, single individuals and two-earner families. In fact, many former Yuppies (Young Upwardly Mobile Professionals) are now becoming Dinks (Double Income, No Kids).

Lengthening life expectancy is increasingly making its own contribution towards the proliferation of family types. There are, for example, three generation families, and—increasingly—four and even five generation families. Three generation families are sometimes referred to as "sandwich" families, since the parents are between the children and the grandparents, and may be caregivers for both. For the older family members, there may be further complications as their children (the parents of their grandchildren) remarry, either via divorce or widow(er)hood. The relationship to the "new" family members; their responsibility for them, in both financial and other terms; and their relationship to the other set or sets of grandparents becomes complicated. Then there are the "second chance" grandparents, whose child, children, or grandchildren come back to live with them after a divorce or a death. Finally, in this area, there is the probability that children, and even grandchildren, will die before the grandparent does. This has been called, "the most distressing and longlasting of all griefs—that for the loss of a grown child."[34]

Sex among the elderly has never been a hot topic of conversation, and efforts to introduce it usually evoke giggles and jokes. But George Bernard Shaw was asked by a reporter on his 80th birthday how it felt to no longer be interested in the other sex. His reply: "I'll let you know." On a more serious note, marriage and remarriage among the elderly is now so common that two consequences have already been remarked.

One is the financial aspect. Pre-nuptial agreements that relieve the anxiety of the children on either side concerning inheritances are common. Further, more and more older couples are living without formalizing the situation into a marriage. Some of this arises from the inheritance question, where laws arbitrarily give the surviving spouse a share of the estate; and some from the pension structure, whereby a couple receives less support than two single people, even if the latter are living together.

The other aspect of marriage among the elderly arises from the phenomenon noted previously, that women tend to outlive men. This means that the number of elderly widowers and divorced men will be much smaller than the number of widows and divorcees. At the end of this century, for every 100 men over age 65, there will be 150 women. Among those over 84, there will be 254 women for every 100 men.[35] Fifty percent of all American women over 65 are widows, and nearly 80% of all married women will become widows.[36]

A constantly growing number of older people, in which women progressively outnumber men, should lead to new societal living arrangements, the structure of which is so far difficult to perceive. Unfortunately, very little research and even less planning seems to be taking place in this area.

CONCLUSION

Among the many changes predicted for the future, the growth of the oldest stratum of the population seems most certain to occur.

The implications of this growth and its characteristics will have enormous consequences for the fabric and structure of society. It is no longer enough for organizations like the United Nations to declare 1999 as the year of the aged. There must be a concerted effort to blueprint future society in which ageism is not seen as a negative phenomenon to be contained, but as an opportunity for a fuller, freer life for most citizens, in which the less fortunate—the ill, the poor, the lonely—are taken care of adequately.

ENDNOTES

1. Cornish, E., "Forecasts Thirty Years Later," *The Futurist*, (January/February, 1997), pp. 45-48.
2. *Demographic Yearbook*, 1996. Geneva: United Nations.
3. *Social Security Programs Throughout the World*, 1995. Washington: US Department of Health and Human Services, 1996.
4. Ibid.
5. Sheldon, S. *Bloodline*. London: Fontana, 1978.
6. Davies, A. M. *Global Aging and Health in the Next Century*. Paper delivered at the 68th International Conference on Social Welfare, Jerusalem, Israel, 1998.
7. World Population Monitoring, New York: United Nations, 1991.
8. Davies, op cit.
9. Schwartz, W. B., "The Conquest of Disease: It's Almost Within Sight." *The Futurist*, 33, 1, January, 1999, pp. 51-55.
10. Davies, Ibid.
11. Demographic Yearbook: Special Issue: Population Ageing and the Situation of Elderly Persons. Geneva: United Nations, 1993, p. 31.
12. *Social Security Bulletin: Annual Statistical Supplement 1996*. Washington: US Department of Health and Human Services, 1997.
13. *Social Policies beyond the 1980s in the European Region*. Vienna: European Centre for Social Welfare Training and Research, 1987.
14. Fowles, D., "The Changing Older Population." *Aging*, 339, pp. 6-11.
15. Research Advances in Aging. 1984-1986. Washington: US Department of Health and Human Services, 1987.
16. Bond, J. "Abstracts: Sociology and Social Policy. *Aging and Society*, 4, 1984, pp. 205-213.
17. Coates, J. F. *The Family in the 21st Century*. Paper delivered at the 68th International Conference on Social Welfare, Jerusalem, Israel, 1998.
18. Zastrow, C. *Social Problems: Issues and Solutions*. Chicago: Nelson Hall, 1992.
19. *Laslett, P. A Fresh Map of Life: The Emergence of the Third Age*. London: Weidenfeld and Nicholson, 1989.
20. Macarov, D. *Certain Change: Social Work Practice in the Future*. Washington: National Association of Social Workers, 1991, p.24.
21. Ibid.

22. Ibid.

23. Kaplan, M. *Leisure: Theory and Policy.* New York: John Wiley and Sons, 1973.

24. Dixon, J., and D. Macarov. *Poverty: A Persistent Global Reality.* London: Routledge, 1998.

25. Macpherson, S., and R. Silburn. "The Meaning and Measurement of Poverty," in Dixon, J., and D. Macarov. *Poverty: A Persistent Global Reality.* London: Routledge, 1998, pp. 1-19.

26. Macarov, D., "Poverty Has a Rich Future," in Didsbury, H. F. Jr. (ed.), *Future Vision: Ideas, Insights and Strategies.* Washington: World Future Society, 1996, pp. 56-75.

27. Whiteford, P., and S. Kennedy. *Incomes and Living Standards of Older People.* London: HMSO, 1995.

28. Chen, Y-P. "Economic Status of the Aged," in Binstock, R. H. & E. Shamas, (eds.), *The Handbook of Aging and the Social Sciences.* New York: Van Nostrand Reinhold, 1985.

29. "Away from Politics," International Herald Tribune, June 19, 1993, p. 3.

30. Hancock, R. "Housing Wealth, Income and Financial Wealth of Older People in Britain," *Aging and Society,* 18, 1998, pp. 5-33.

31. Gal, J. The Changing Relationship between Work and Social Security –Reflections on Future Trends. Paper delivered at the 68th International Conference on Social Welfare, Jerusalem, Israel, 1998.

32. Social Security Programs Throughout the World, 1997. Washington: US Department of Health and Human Welfare, 1998.

33. Wellness Letter, 1989, 5, 1.

34. Gorer, G., Death. *Grief and Mourning.* New York: Doubleday, 1965.

35. Hartman, A. "Aging as a Feminist Issue." *Social Work,* 35, 5, 1990, pp. 387-388.

36. Kamm, P. S. *Remarriage in the Middle Years and Beyond.* San Leandro, CA.: Bristol, 1991.

COMPSPEAK 2050: HOW TALKING COMPUTERS WILL RECREATE AN ORAL CULTURE BY MID-21st CENTURY

by

William Crossman

(This article is a slightly revised version of Chapter 1 of the author's forthcoming book, also entitled *CompSpeak 2050: How Talking Computers Will Recreate an Oral Culture by Mid-21st Century*.)

The 20th Century is behind us, the new millennium has arrived, and, in the United States:

- most people would rather talk to someone on the telephone than write to them;
- most people would rather watch TV than read a book;
- most schools and most school children are engulfed in a deep literacy crisis, with little hope for a breakthrough.

Why is all this happening now? Superficial explanations, such as blaming it on people's mental laziness or backwardness, on TV, or on the schools, simply won't do. Something much deeper, and more difficult to see, is going on.

The growing feelings of alienation from writing and reading, which school children and people of all ages are experiencing and expressing through their day-to-day behavior, are signs and symptoms of a profound historical, social, technological, and evolutionary change. They are symptomatic of a massive shift that is taking place: away from the use of written language and back to the use of spoken language to communicate, store, and retrieve information in our daily lives.

In the United States and other electronically-developed countries, we're witnessing nothing less than the abandonment of reading and writing, of written language itself, and, in its place, the recreation of oral culture. The push to develop voice-recognition technology and VIVOs—Voice-In/Voice-Out computers that we can talk to and that can talk back to us—is part of this evolutionary leap.

It truly is evolutionary. Historically, before our human ancestors developed written language, they accessed stored information orally-aurally, by speaking and listening—as well as by seeing, smelling, tasting, and touching. They relied on their memories to store information that they heard—as well as saw, smelled, tasted, and touched—and they retrieved it for others by speaking and acting.

William Crossman *is a professor, speaker, and consultant at CompSpeak 2050 Instutite, Oakland California.*

Six thousand to 10,000 years ago, people's memories were no longerefficient and reliable enough to store and retrieve the influx of new information that arose with the onset of the agricultural revolution. To transcend their memories' limits, our ancestors came up with a remarkable solution: written language.

It was a feat that required great imagination and complexity of thought and, in today's terms, involved the creation of: new software—pictographs and alphabets forming written languages; new hardware or the adaptation of old hardware to new tasks—pens, pencils, brushes, knives, inks, chalk, pigments, animal skins, paper, leaves, wood, stone; and new operations—writing and reading. Using this new technology, our ancestors freed themselves from the limits of human memory. Written language enabled them to freeze and thaw as much information as their hardware allowed.

Today, we in the electronically-developed countries view writing and reading as one of the necessities of human existence, as something we can't do without, like water, food, and sleep. This may be the view we see through our culturally-biased, pro-text eyeglasses, but it's just plain wrong. Not only is written language not necessary to human existence, but we could have reached today's level of information storage-retrieval without ever having created written language in the first place.

If some early society had found or carved a wooden, bone, or stone cylinder, coated it with beeswax, attached a porcupine quill to a hollow gourd, let the quill rest on the wax-coated cylinder, and spoken into the gourd while rotating the cylinder, written language might never have happened. Humanity might have gone right from storing and retrieving speech-based information by memory to doing it by phonograph without entering the world of print culture at all. Humanity's First Golden Age of oral culture might never have ended.

I'm overdramatizing this point somewhat to help lay the groundwork for a different view of written language. Throughout this article, I characterize written language as a technology, a technological solution to specific information-storage-retrieval problems that people faced at a specific moment in history 6,000 to 10,000 years ago. Like most technologies, written language will serve its function until some better technology comes along to replace it. Written language isn't an eternal verity. We can admire it, but we shouldn't worship it.

With written language about to make its exit, and its replacement already stepping through our front door, it is vital that we see written language clearly for what it is: a transitory technology. This reality-check will help us prepare ourselves to say "goodbye" to it and to welcome back its replacement: our old friend, spoken language.

Unlike written language, spoken language—by which I mean speech itself—wasn't/isn't a technology devised by people to overcome human limitations in the face of social and environmental changes. In this sense, spoken language isn't a technology at all. Though we humans created or devised particular spoken languages, we didn't create or devise spoken language itself any more than we created our circulatory systems.

Our ability to speak language, period, is an inborn characteristic of our species. We carry in our genes and our brains the capacity for spoken language. If the day ever arrives that we wave a final "good-bye" to spoken language—and to the sign languages used by people with hearing and/or speaking disabilities—we'll be waving "good-bye" to the species of human beings that we are.

In contrast to written language, spoken and sign language is "user friendly." As very young children, we just start understanding it and speaking or signing it. We don't have to spend years in school learning to speak. Nor does spoken language drive a wedge into the world's population the way written language does—dividing humanity into those who can read and write and those who can't. Everyone who is mentally and physically able can speak a language.

Historically, spoken language came—and had to come—to humans before written language. Biologically, speech or sign language comes—and has to come—to each child before literacy. This is because written languages are symbolic representations of spoken languages. Had we no spoken language, we could not have created written language. Written language may have emerged as the primary method used to store and retrieve information in certain areas of the world, but it is based on and derived from spoken language.

In the 21st Century, people with access to VIVO-computer technology will once again be able to use spoken language to access all stored information. Talking computers are going to make writing, reading, spelling, alphabets, punctuation, written numerals, music notation, and all other notational systems obsolete.

The obituary for written language won't be written. It will be spoken by someone talking to a VIVO computer in 2050.

Since the mid-19th Century, humanity has been waging a furious assault against written language. This has taken the form of people—particularly North Americans and Europeans—inventing and developing devices which use spoken language, rather than written language, to communicate, store, and retrieve information. A cornucopia of speech-based devices now exists which has simplified and sped up—and/or completely redefined—the tasks we formerly assigned to text and text-based devices. For the past 125 years, in North America, Europe, and Japan, these speech-based devices have been relentlessly usurping the functions of the text-based devices.

The letter, the magazine, the newspaper, the broadside, the book, the flyer, the written advertisement, the memo or written message, the file, the written record, the official document, the written school exercise—all have come under attack. In some cases, direct or instantaneous oral-aural communication devices (telephone, telegraph, live radio, live television) have been doing the usurping. In other cases, devices which store information in the form of speech (phonograph, audiocassette, "talkie" movie film, videocassette) have been responsible.

Since 1990, our minds and resources have turned to the development of talking computers. In our rush to create VIVOs, we're continuing the process Edison, Morse, Tesla, Bell, and their counterparts began.

E-mail appears to be an exception to the above: a form of written message whose popularity is surging in the electronically-developed countries. I predict, however, that the moment we're linked together by VIVOs—a moment merely several years away—most of us will stop typing our messages and will start speaking them again.

Why have we been so obsessed with researching and developing oral-aural replacements for written language? Because biologically, psychologically, and technologically, we have again hit limits on the efficiency and reliability of our main method for freezing and thawing information. Formerly, as I mentioned above, it was the limits of human memory to retain the influx of information during the agricultural revolution that led people to create written language. For the last 125 years, it has been the *limits of written language use* that have driven people to seek and develop oral-aural replacements. I'll return to these limits in a moment.

Even though the scientific sector has been working overtime these past 125 years to develop oral-aural and non-text visual technologies, from the wax-cylinder phonograph to the talking computer, the true nature of this process and its goal—to supersede written language's limits by returning to oral-aural methods of information communication and storage—has been largely *undeclared, unacknowledged, and unconscious,* even among the chief developers themselves.

This article has, as one of its main objectives, to acknowledge this process and to raise it to the level of consciousness and awareness. If we understand what is happening and why, we'll be better able to evaluate it and direct its course.

VIVOs will be the last nail in written language's coffin. By making it possible for us to access stored information orally-aurally, talking computers will finally make it possible for us to replace all written language with spoken language. Once again, we'll be able to store and retrieve information simply by talking and listening—and by looking, too, but at graphics, not at text. With this giant step forward into the past, we're about to recreate oral culture on a more efficient and reliable technological foundation.

From a Darwinian perspective, written language is a 6,000-to-10,000-year-old bridge that humanity has been using to walk from our First Golden Age of oral culture to our Second. We undertook this journey to survive as a species. Six thousand to 10,000 years ago, lacking the ability to store and retrieve by memory the growing sum of survival information, our species faced two options: develop new storage-retrieval technology or self-destruct. That's when and why we created the written-language bridge.

As a species, humans have instinctively understood that any systematic failure in our ability to store and retrieve information is a threat to our survival. Now, we in the print-literate nations are instinctively reacting to the fact that written language—our stored-information accessor of choice—has hit its limits and is failing us.

It is failing us, first, because it is no longer able to do the tasks we created it to do and, second, because too many people are unable to use it. An example of the first: for most literate people, communicating, storing, and retrieving information by writing and reading is still far slower and more tedious than doing it by speaking.

Regarding the second, it's sufficient to remind ourselves that the great majority of the world's people, by conservative estimates 80% of humanity, including many living in the so-called print-literate nations, still can't use written language effectively. Most societies in the world today are still oral cultures, and very few of the world's societies—including the United States—possess either the enormous human and economic resources and/or the political will required to fully train their populations to write and read.

There's another reason—also related to evolution and natural selection—that we're leaving written language behind. In carrying out the historical commitment that we, in the print-literate societies, had made to written language, we have—largely unaware of our dangerous path—strayed too far away from our innate information-communication-storage-retrieval method: speech.

Our genes, nervous systems, muscles, and emotions have been sending us a crisis wake-up call, reminding us of our spoken-language survival mandate and telling us to return to our oral-aural roots, or else. Or else what? Or else the speech-deficiency-based physical and mental illnesses—similar in many ways to sun-, motion-, sleep-, and vitamin-deficiency illnesses—that began to strike the print-literate nations in the 19th Century and that have become an epidemic in the late-20th Century, will continue to spread unchecked.

Since the late-1800s, we in the print-literate nations have been acting swiftly and positively—though mainly unconsciously—to avert this health crisis. We've been heeding the evolutionary mandate of our human physiology and psychology to reverse the widening gap between our present print-oriented selves and our innate, biogenetic, oral-aural selves. It seems clear to me now that the steps we have been taking to stem this epidemic include phasing out writing and reading entirely and phasing in speech-based devices including talking computers.

On some unconscious mental level, we seem to understand or believe that talking computers will help us to achieve the wellness and wholeness we seek. We seem to unconsciously understand that hooking ourselves up to these talking-computer I.V. units—IntraVIV-Os?—will rid us of textual toxicity and pump us full of lifesaving orality. I am overdramatizing again, but only a bit.

Located at the far end of the written-language bridge, the Second Golden Age of oral culture has been visible to us since the invention of the phonograph in 1877. We've trekked the bridge toward our destination decade after decade through the 20th Century. By mid-21st Century, we will have finally reached the bridge's end and stepped off into the future. Once across, we will never look back.

School children's declining literacy rate is a symptom of these deeper processes. As a group, young people in the electronically-developed countries have chosen oral-aural and non-print visual technologies—video, stereo, radio, film, telephone, and computer—as their preferred methods for accessing "live" and stored information.

These technologies, like written language, are external extensions of our brains' memory banks and our sense organs—mindparts located outside of our heads. But unlike written language, they allow us to communicate in the way that's most basic and familiar to us: through spoken language.

Most young people today instinctively understand this rock-bottomness of speech/spoken language. They are in touch with it. They feel it in their bones, their brains, their genes. Why should they read and write, so many young people ask, when they can listen and speak? They view the rules of writing as they view all rules imposed on them by adult society—as devices to dominate and control young people. And they're rebelling.

Students' refusal to go along with the program is causing our schools to develop a record of failure, as each twelfth-, eighth-, or third-grade class graduates with a weaker grasp of reading and writing than the prior year's twelfth-, eighth-, or third-grade class. Writing teachers are feeling discouraged and demoralized, and many have basically given up trying to teach it. The result: a downward spiral of writing-reading skills and test scores that has become the school literacy crisis of the 1990s.

By 2050, if large numbers of students have been able to gain access to talking computers, all this negativity and failure concerning writing and reading will be a distant memory. All education in the electronically-developed countries will be oral-aural and non-text visual. Students will use talking computers with optional monitors displaying icons, graphics, and visuals to freeze and thaw information.

Instead of the "three R's"—reading, 'riting, and 'rithmetic—students will focus on the "four C's"—critical thinking, creative thinking, compspeak, and calculators. I call it VIVOlutionary learning.

We won't have to wait until 2050. By *2005*, a student assigned to write an essay will be able to speak it into a VIVO computer, use VIVO's grammar-check to organize and correct it, "proofread" it by listening to VIVO repeat it, print it out, and submit it to the teacher for a grade.

The student will have proven two things. First, that any person—nonliterate as well as literate—with a talking computer will be able to produce a perfect written essay. Second, that *because* any person with a talking computer will be able to produce a perfect written essay, written language will have become obsolete.

Why should the student in the above example bother to print out a copy of the essay? Why should they bother with that final step of translating their spoken ideas into written language? Their teacher certainly doesn't need a written record of their ideas. Using their own VIVO, the teacher will be able to listen to the student's spoken ideas online and respond accordingly. Neither the student nor the teacher needs to write anything down in order for learning to occur and for education to take place.

In this scenario, the student and teacher are using their VIVOs exactly the way VIVOs are supposed to be used. Isn't that why we're developing VIVOs? Isn't that what they're for? Don't we *want* students to be able to input their ideas orally online and teachers to be able to access those ideas aurally? Voice-in, voice-out: simple.

We developed written language to store and retrieve information, and we are developing talking computers to perform the very same function. Because talking computers will do it more easily, quickly, efficiently, universally, and (ultimately) cheaply, they will replace written language. Simple.

We used to cut our grass with a scythe; then, we invented the push lawn mower and put the scythe in a museum; then, we invented the gasoline-engine lawn mower and put the push lawn mower in the museum. That's the way technology works, and the way we work with technology: we are forever replacing the old with the new.

In the case of written language, however, we are replacing a technology (written language) with a non-technology (spoken language), but we are giving the non-technology a new technological twist: an electronic echo, a gigantic memory capable of storing and retrieving an almost unlimited amount of information in the form of speech.

Written language was a technology created by our ancestors to help them deal with a specific set of historical needs and conditions in a specific historical period several thousand years ago. Today, we are creating VIVO technology to answer a different set of needs and conditions in our own historical period. Soon we'll be placing written language on the museum wall next to the scythe.

Just as some students today might join a choral group, karate club, or chess club as a pleasurable pastime, some mid-21st-Century students might join a literacy club to learn written language for fun. But there will be no compelling reason why they would need to learn to read and write and, therefore, no compelling reason why they should have to learn it—or why their schools should have to teach it. Exit the school literacy crisis.

Not only education but the arts and, possibly, international relations will be transformed in the shift from print to oral culture.

Imagine the literary arts without written language, and the musical arts without written music: a return to story*telling*, spoken poetry, and improvised music.

Imagine international relations without written language: dominant nations would no longer be able to force other nations to read and write in the former's "standard" languages—a traditional weapon of cultural domination—and would no longer be able to decide which individuals, in the dominated nations, would be allowed to become literate.

These are just two examples of areas in which VIVOs, or more accurately, people using VIVOs, will reshape the world in the 21st Century. In this article, I take the viewpoint that *good* results *could possibly* come from the fact that talking computers will soon take over written language's job. Lovers of the written word—and I am one of you—I invite you to give the following ideas a hearing.

The creation of VIVOs will create new *potential opportunities* for people in three areas.

• VIVOs will create new *potential opportunities* for the world's nonliterate and semi-literate people to be able to access—through speech or signing alone—the world's storehouse of information and knowledge. For the first time since the introduction of written language, people's nonliteracy or semi-literacy won't prevent their accessing all stored information.

Pre-VIVO electronic technologies have already actualized similar potential opportunities for millions, maybe billions, of people worldwide. Within a period of about 60 years, a huge amount of information that had been formerly inaccessible, because it had been stored in the form of written language, has become available to people who can't write or read. Radio, video, stereo, film, telephone, and computer have opened up an oral-aural and/or non-text visual universe of stored information for the non-readers and non-writers who have finally been able to gain access to these technologies.

Sixty years isn't a long time. The very existence of written language on Earth for sixty-hundred years or more has profoundly affected and reshaped all cultures and communities—even those that are still oral cultures. Now, even before VIVOs sprout from our wrists and lapels, radio, video, and the rest have been busily, and irreversibly, reshaping global reality once more.

• VIVOs will create new *potential opportunities* for all people, whether literate or not, to instantaneously communicate *in all languages* with other speakers or information storage units. Using today's "old-fashioned" text-entry computers and text-translator software, a person can communicate in writing with another person who reads and writes another language. The problem is that both people must already be able to read, write, and enter text in at least one language, their own. Using VIVOs, we won't need to know how to read or write, won't need to be verbally fluent in any language other than our own native language, and won't even need to understand a universal language like Esperanto. VIVO units will allow people around the world to speak easily with one another in their respective native languages, thanks to VIVO's simultaneous speech translation function. Electronic Esperanto!

• VIVOs will create new *potential opportunities* to access stored information for many people who can speak and hear, or sign and see, but whose physical and/or mental disabilities make it difficult or impossible for them to write and/or read.

In these last three bulleted sections, I've italicized the words "potential opportunity" with good reason. The most that *the birth* of a new technology can possibly achieve is to open potential opportunities for people. These opportunities can only become actualized when people actually gain access to the technology and utilize it.

Here's a familiar example. Since most people in the world can speak and hear, they have the ability to use the telephone. But most of the world's people don't have access to telephones because of the high cost of service and/or the unavailability of service in their communities. The invention of the telephone has opened up the *potential opportunity* for everyone to speak across long distances, but using a telephone remains out of reach for the billions of people who haven't been able to get their hands on one.

The birth of a new technology, by itself, can't change anything. People having access to and using the new technology can create change and make history.

For the foreseeable future, which includes the VIVO Age of the 21st Century, the issue will continue to be: who controls the new technology and, therefore, who controls whom. As with all technology, talking-computer hardware and software will be developed, patented, copyrighted, programmed, manufactured, encrypted, sold, bought, leased, used, distributed, and shared—or not shared—by those with the wealth and resources to control these processes.

In my opinion, it would be great if all the nonliterate and semi-literate people around the world could actually start using VIVOs on January 1, 2010 to access the world's databanks. But it would be naive to think this will happen automatically just because the technology itself will exist on January 1, 2010. If people want access to talking computers, if we want to actualize the potential opportunities VIVO technology presents to us, we've got to figure out how to do it...and then do it. The right to have access to the stored information, the collective knowledge, of our community, our society, and our world is a human right. The ability to read and write, print-literacy, is still the key that opens these information vaults. Yet, billions of people around the world are being denied access to this information because they remain nonliterate or semi-literate. Most of the world's people still haven't received their keys. Literacy has historically been treated as a privilege, rather than a right, by those who hold the master keys—and it's as true today as ever.

Most people in the world haven't even seen a library. If they were able to travel to a library, and if they were able to find its door open to them—for many libraries' doors would not be—they would still find the meanings of the sentences in the books on the library's shelves closed to them.

In this article, I say some negative-sounding things about writing and reading. For example, I say that they are about to become obsolete. This isn't intended as an attack on written language. It's merely an observation, part of a broader analysis presented here. And it's definitely not intended as an attack on, or a demeaning of, people worldwide who are striving to learn to write and read.

I have the greatest respect and admiration for children and adults around the world who are trying to achieve literacy. Today, a person's ability to write and read still raises for them new possibilities of communication, knowledge, social and political involvement, employment, enjoyment, self-esteem, creative expression, and much more.

The analysis presented here, rather than closing the doors of hope on those who lack and/or seek the literacy-key to information, communication, and knowledge, opens these doors wide for them. It says that the same VIVO-computer technology that will make written language obsolete will also create potential opportunities for the billions of nonliterate and semi-literate people throughout the world to tap into the world's store of information—without having to learn to read and write at all.

To all students and all others who wish to read and write, I still say: Go to school! Stay in school! Learn to read and write! We'll need it in our lifetimes!

I say this because written language isn't going to disappear overnight. It will take time: decades, two or three generations, maybe even a whole century. Yet, its eventual disappearance will be the end of a process that's already well underway.

It is ironic that, with the 21st Century's arrival, oral culture is growing in the print-saturated, electronically-developed countries at the exact moment that many electronically-undeveloped countries are earnestly launching literacy campaigns. The near-future course of literacy development in these latter countries is very difficult to predict. However, in the United States today, we need only to look around to confirm that a massive, rapid, electronically-aided decline in the number of writers-readers has begun.

We're witnessing the beginning of an earthshaking transformation of human society away from print culture toward oral culture. It will occupy and involve the energies of humankind from the beginning to the end of the 21st Century and will surely reshape every field of human activity. By 2050, in the United States and the other electronically-developed countries, the use of written language, of writing and reading, will mainly be a thing of the past.

One hundred years beyond, by 2150, all of the world's communities will again be oral cultures. VIVOs will link all of them—including those we build in space—into a single, oral-aural, information-access network: a true, worldwide oral culture.

THE GLOBAL IMPACT OF INFORMATION TECHNOLOGY: THE CONNECTED VERSUS THE UNCONNECTED

by

John W. McDonald

The World is not structured today to cope with the dramatic global challenges presented by the information technology revolution.

The United States, and the rest of the developed world, have already outstripped the developing world, in access to and use of the Internet, as it grows exponentially, before our very eyes.

It is exciting to learn there are 100 million Internet users, linked to some 30 million computers around the world. It is shocking to be told however, that more than half of these users are in the United States and that of the 100 million users, 92 million are in the developed world. That means that among the remaining 5 billion people in the world, there are 8 million people who have access to the Internet!

Unless steps are taken now, to identify and acknowledge this growing disparity between the haves and the have-nots, between the developed and the developing world, between the rich and the poor, the rural and the urban, the literate and the illiterate, the English speakers and the non-English speakers, the long-term, negative repercussions that will permeate our global society, will be enormous. This ever widening gulf poses a threat to our hopes and aspirations for a more democratic world. If this threat is not recognized, and we continue down our present path, the impact of the IT revolution will be enormously destructive to the lives of most of the people on the planet.

New international approaches must be designed to help that part of the world that does not have access to this new technology. If this does not happen and the world continues to ignore the current situation, in a single generation, two-thirds of the world, by that time, 7 billion people, will be politically, economically and culturally disenfranchised and further impoverished.

This is a major challenge for the 21st century.

We can learn from the past.

In the years following World War II, after much of Europe had been re-built and the world's great empires had begun to collapse, spawning dozens of new, impoverished, developing countries, a few leaders began to realize that the world was being divided, economi-

John W. McDonald *is chairman and co-founder of the Institute for Multi-Track Diplomacy in Washington, D.C. Ambassador McDonald retired from the foreign service in 1987.*

cally, between the North and the South, between the "haves" and the "have nots."

The leaders were from the "Third World," the developing world, not from the North.

These leaders, like Dr. Raoul Prebish, from Argentina, decided to take action themselves and do something about this inequity, much to the chagrin and the concern of those in the North.

Something called the "North-South Dialogue" began to take shape.

In 1962, Dr. Prebish asked the United Nations General Assembly to approve the calling of the first World Conference on Trade and Development. The General Assembly agreed and the Conference took place in Geneva, Switzerland, in 1964. In his report back to the General Assembly that year, on the results of this global meeting, Conference Secretary General Prebish was astutely able to turn the "Conference" into a new United Nations Organization, known as UNCTAD, which continues its work to this day. It was the first time the developing world had a platform from which to speak about its concerns and frustrations, and be heard by the North. For the first time the Third World had a voice at the negotiating table, and they used it most effectively.

In fact, this model was so successful that it was used a second time in 1966. The developing world was able to get General Assembly approval to create the United Nations Industrial Development Organization, UNIDO, based in Vienna, Austria. It was led by a distinguished Egyptian, Dr. Abdul Rahman. UNIDO's mission was to encourage the West to help the Third World create small business opportunities in the developing world. Again, the South had a voice at the negotiating table.

These two global organizations, dominated by the South, forced Northern governments to officially recognize that inequities did exist between the North and the South, in the areas of trade and development and small business. Northern governments were obliged to focus on these development issues. The struggle between the North and the South still goes on but these two United Nations organizations have made an enormous difference, over the years, in the way the North evaluates and responds to the global problems faced by the developing world.

The World would be a far worse place to live in today, if dynamic institutional action had not been taken in the mid-1960's, sparked by leaders from the Third World.

Everyone agrees that access to information technology must begin with education. How many of us realize, however, that over a billion people in the world are illiterate, and most of those are women and children, especially girl children?

Over three billion more people have no access to computer technology because they have no access to electricity, or because they are financially not able to afford to buy a computer. Then there are hundreds of millions of others who, for domestic political reasons, are denied unobstructed access to this new technology.

What can be done about this situation - today?

The World Bank, under the vigorous leadership of its president, Mr. Jim Wolfensohn, took a major first step in 1995, at the Fourth United Nations Conference on Women and Development, in Peking. He pledged to bring a sixth grade education to all girl children and boy children in the world, by the year 2015!

On his return to Washington, President Wolfensohn began to make good on his personal commitment to the World. He convened all of the relevant United Nations Agencies and, with financial support from The World Bank, got them to agree to his goals. This system-wide, joint attack on illiteracy, is a major first step to help the next generation begin the process of eventually having access to IT.

This global approach has to be expanded, today, by taking additional steps, at many levels, if we want to make information technology available to all people, across the world.

It is imperative that the developed countries not wait, as they did in the 1960's, and then be forced to take emergency action. The United States private sector and the US government should take the lead in alleviating these current and future inequities in access to IT. The unconnected must become the connected.

Our rapid technological advances will soon isolate us from all but the elites of the world. Over time this technological isolation will lead to sociological, political and economic instability, and eventually violence. It is politically untenable for the world's greatest democracy, and only superpower, to be seen as standing by and doing nothing.

Leading figures in the IT private sector community in the United States, should demonstrate a real commitment to the resolution of the problem and should step forward. The richest man in the world, Mr. Bill Gates, could offer to put a computer in every Library in the world, (Senator Al Gore's idea), and link them to the Internet at an estimated cost of $300 million. What a global breakthrough that would be!

The terms of reference for Ted Turner's generous gift to the United Nations, of $100 million a year for ten years, could be amended to allow a new structure to be created in the United Nations system, to focus on this issue.

On the governmental side the United States could strongly urge United Nations Secretary General, Mr. Kofi Annan, to convene a small, high-level, informal meeting, of concerned Government representatives, UN Agency representatives, private IT business leaders and non-governmental organizations, to discuss these issues. As consensus evolves on the seriousness of the problem, this group could begin to focus on specific next steps that could be taken, by working together.

One step they might take would be to recommend a lead UN Agency, such as the International Telecommunications Union, to coordinate action within the UN system.

A second step would be to urge the creation of a new United Nations Commission, under the aegis of the UN Economic and Social Council, to be called the "International Technology Commission."

This IT Commission would operate across formal and informal boundaries, by bringing together representatives from governments, the UN, non-governmental organizations and the private sector, who currently do not even talk to each other, and get them to learn how to work together. Their collective wisdom, under this new and innovative structure, could begin to cope with this crisis in information technology for all.

Action taken now will dramatically reduce future political power distortions, conflict and violence in the early part of the next century. Those without hope of access to the new information technology, under our current structure, may soon decide to fight for that right of access. Action today will prevent that from happening.

PROSPECTS FOR GLOBAL FOOD SECURITY IN THE TWENTY-FIRST CENTURY

by

Per Pinstrup-Andersen and Rajul Pandya-Lorch

INTRODUCTION

Dramatic transformations have occurred in recent decades in where and how food is produced, processed, and traded such that enough food is now available to meet the basic needs of each and every person in the world. The doubling of grain production and tripling of livestock production since the early 1960s has resulted in about 2,700 calories available per person per day. However, about 820 million people lack access to sufficient food to lead healthy and productive lives and around 185 million children are seriously underweight for their age. At the close of the twentieth century, astonishing advances in agricultural productivity and human ingenuity have not yet translated into a world free of hunger and malnutrition.

OUTLOOK FOR GLOBAL FOOD SECURITY

Projections of food production and consumption to the year 2020 offer some signs of progress, but prospects of a food-secure world—a world in which each and every person is assured of access at all times to the food required to lead a healthy and productive life—remain bleak if the global community continues with business as usual. IFPRI's global model, the International Model for Policy Analysis of Commodities and Trade (IMPACT), projects that under the most likely or baseline scenario, 150 million children under the age of six years will be malnourished in 2020, just 20% fewer than in 1993 (Figure 1). One out of every four children will be malnourished in 2020, down from 33% in 1993. Child malnutrition is expected to decline in all major developing regions except Sub-Saharan Africa, where the number of malnourished children could increase by 45% between 1993 and 2020 to reach 40 million. In South Asia, home to half of the world's malnourished children in 1993, the number of malnourished children is projected to decline by more than 30 million between 1993 and 2020, but the incidence of malnutrition is so high

Per Pinstrup-Andersen, *a native of Denmark, is the director general of the International Food Policy Research Institute, which is part of a global agricultural research network, the Consultative Group on International Agricultural Research.* **Rajul Pandya-Lorch,** *a citizen of Kenya, is head of IFPRI's 2020 Vision for Food, Agriculture, and the Environment, an international initiative that seeks to identify solutions for feeding the world, alleviating poverty, and protecting natural resources.*

that, even with this reduction, two out of five children could remain malnourished in 2020. With more than 70% of the world's malnourished children, Sub-Saharan Africa and South Asia are expected to remain "hot spots" of child malnutrition in 2020.

FIGURE 1

NUMBER OF MALNOURISHED CHILDREN, 1993 AND 2020

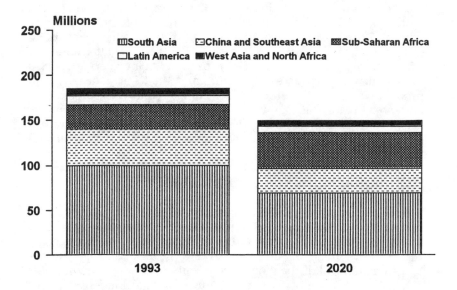

Source: IFPRI IMPACT simulations.

IFPRI projects global demand for cereals to increase by 41% between 1993 and 2020 to reach 2,490 million metric tons, for meat demand to increase by 63% to 306 million tons, and for roots and tubers demand to increase by 40% to 855 million tons. Developing countries will drive increases in world food demand. With an expected 40% population increase and an average annual income growth rate of 4.3%, developing countries are projected to account for more than 85% of the increase in global demand for cereals and meat products between 1993 and 2020. Nevertheless, disparities in consumption will remain wide between the developed and developing world: a developing-country person is forecast to consume, on average, only 40% of the cereals and 35% of the meat products that a developed country person would consume in 2020.

How will the expected increases in cereal demand be met? Not by expansion in cultivated area. The burden of meeting increased demand for cereal rests on improvements in crop yields. However, the annual increase in yields of the major cereals is projected to slow down during 1993–2020 in both developed and developing countries. This is worrisome given that yield growth rates were already on the decline. With the projected slowdowns in area expansion and yield growth, cereal production in developing countries as a group is also forecast to slow to an annual rate of 1.5% during 1993–2020 compared with 2.3% during 1982–94. This figure is still higher, however, than the 1.0% annual rate of growth projected for developed countries during 1993–2020.

Food production will not keep pace with demand in developing countries and an increasing portion of the developing world's food consumption will have to be met by imports from the developed world. The proportion of cereal demand that is met through net imports is projected to rise from 9% in 1993 to 14% in 2020. As a group, developing countries are projected to more than double their net imports of cereals (the difference between demand and production) between 1993 and 2020 (Figure 2). With the exception of Latin America, all major developing regions are projected to increase their net cereal imports: the quadrupling of Asia's net imports will be driven primarily by rapid income growth, while the 150% increase forecast for Sub-Saharan Africa will be driven primarily by its continued poor performance in food production. The United States is forecast to provide almost 60% of the cereal net imports of developing countries in 2020, the European Union about 16%, and Australia about 10%.

With continued population growth, rapid income growth, and changes in lifestyles, demand for meat is projected to increase by 2.8% per year during 1993–2020 in developing countries and by 0.5% per year in developed countries. While per capita demand for cereals is projected to increase by only 8%, demand for meat will increase by 43%. The increase in per capita meat demand will be largest in China and smallest in South Asia; by 2020, Chinese per capita consumption of meat will be eight times that of South Asia. Meat production is expected to grow by 2.7% per year in developing countries during 1993–2020 (compared with 5.9% during 1982–94) and by 0.8% in developed countries (compared with 0.9% during 1982–94). Despite high rates of production growth, developing countries as a group are projected to increase their net meat imports 20-fold, reaching 11.5 million tons in 2020. Latin America will continue to be a net exporter of meat, but Asia will switch from being a small net exporter to a large net importer.

Net imports are a reflection of the gap between production and market demand. For many of the poor, the gap between food production and human needs is likely to be even wider than that between production and demand, because many of these people are priced out of the market, even at low food prices, and are unable to exercise their demand for needed food. The higher-income develop

FIGURE 2

NET CEREAL IMPORTS OF MAJOR DEVELOPING REGIONS, 1993 AND 2020

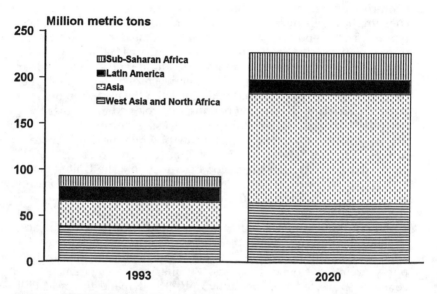

Source: IFPRI IMPACT simulations.

ing countries, notably those of East Asia, will be able to fill the gap between production and demand through commercial imports, but the poorer countries may be forced to allocate foreign exchange to other uses and thus might not be able to import food in needed quantities. It is the latter group of countries, including most of those in Sub-Saharan Africa and some in Asia, that will remain a challenge and require special assistance to avert widespread hunger and malnutrition.

EMERGING ISSUES IN GLOBAL FOOD SECURITY

IFPRI's projections represent the likely future food situation, but how closely these projections will match reality depends on a number of factors that influence food supply and demand. These factors suggest that humanity is entering an era of greater fluctuation and risk in the world food situation.

- Grain prices may be more volatile in the future even though the long-term trend for cereal prices continues to decline.
- Policy decisions, as well as changes in lifestyles and income levels, in China and India—the two most populous countries in the world, with rapidly transforming economies—can affect food security not only for large populations in these countries themselves, but also for the rest of the world.
- With the political changes in Eastern Europe and the former Soviet Union, many projected that these countries would quickly switch from being net importers of grains to being significant net exporters. Although this scenario has not materialized, IFPRI projections show that with appropriate reforms this region could become a substantial net exporter.
- Sub-Saharan Africa is experiencing a fragile economic recovery, but growth rates will have to be substantially higher if they are to make a dent in the region's poverty. More will have to be done to support farmers and reduce population growth if economic growth is to be sustained and food security improved.
- The United Nations recently scaled back its population projections, but the world's population is still projected to increase by 35% during the next quarter century, from 5.69 billion in 1995 to 7.67 billion by 2020.
- Weather fluctuations could affect food production in many parts of the world. The resurgence of El Niño is causing widespread weather anomalies, and the long-term effects of climate change on food production are still unclear.
- The world is placing increasing demands on its fixed supply of renewable water resources. Unless properly managed, fresh water may well emerge as the key constraint to global food production.
- Improved soil fertility is essential for low-income countries to increase sustainable agricultural production. Fertilizer use in many areas will have to grow much faster than projected if farmers are to produce enough food and conserve natural resources.
- Food aid and other international development finance to developing countries are falling, threatening poor countries' ability to cope with food emergencies and achieve

food security.

- Although many developing countries have opened their markets to foreign trade, many developed countries have not, preventing developing countries from fully benefiting from trade liberalization.
- The movement in North America and Western Europe toward limiting some use of genetic engineering and other tools of modern science for food production and processing could threaten food supplies in developing countries, partly because of reduced exports by developed countries and partly because developing countries might adopt similar policies, constraining agricultural production.

REQUIRED ACTION TO ASSURE GLOBAL FOOD SECURITY

- Strengthen the capacity of developing-country governments to perform appropriate functions, such as maintaining law and order, establishing and enforcing property rights, promoting and assuring private-sector competition in markets, and maintaining appropriate macroeconomic environments.
- Invest more in poor people in order to enhance their productivity, health, and nutrition and to increase their access to remunerative employment and productive assets. Governments, local communities, and NGOs should assure access to and support for a complete primary education for all children; assure access to primary health care for all people; improve access to clean water and sanitation services; provide training for skill development in adults; and strengthen and enforce legislation and provide incentives for empowerment of women to gain gender equality.
- Accelerate agricultural productivity. The key role of the agriculture sector in meeting food needs and fostering broad-based economic growth and development must be recognized and exploited. To make this happen, agricultural research systems must be mobilized to develop improved agricultural technologies, and extension systems must be strengthened to disseminate improved technologies.
- Promote sustainable agricultural intensification and assure sound management of natural resources. Public- and private-sector investments in infrastructure, market development, natural resource conservation, soil improvements, primary education and health care, and agricultural research must be expanded in areas with significant agricultural potential, fragile soils, and large concentrations of poverty to effectively address their problems of poverty, food insecurity, and natural resource degradation before they worsen or spill over into

other regions.

- Develop effective, efficient, and low-cost agricultural input and output markets. Governments should phase out inefficient state-run firms in agricultural input and output markets and create an environment conducive to effective competition among private agents in order to provide efficient and effective services to producers and consumers.
- Expand and realign international assistance and improve its efficiency and effectiveness. The current downward trend in international development assistance must be reversed, and industrialized countries allocating less than the United Nations target of 0.7% of their gross national product (GDP) should rapidly move to that target. International development assistance must be realigned to low-income developing countries, primarily in Sub-Saharan Africa and South Asia where the potential for further deterioration of food security and degradation of natural resources is considerable.

CONCLUSIONS

Food insecurity has long been perceived by some to be primarily a problem of insufficient food production rather than insufficient access to food. Yet, as enough food is being produced to meet the basic needs of every person in the world, it is evident that the persistence of food insecurity—about 820 million chronically undernourished people and 185 million malnourished children—is increasingly attributable to difficulties in accessing sufficient food. Food-insecure people simply do not have the means to grow and/or purchase the needed food. Empowering every individual to have access to remunerative employment, to productive assets such as land and capital, and to productivity-enhancing resources such as appropriate technology, credit, education, and health care is essential. Besides enabling every person to acquire the means to grow and/or purchase sufficient food to lead healthy and productive lives, assuring a food-secure world calls for producing enough food to meet increasing and changing food needs and for meeting food needs from better management of natural resources.

With foresight and decisive action, we can create the conditions that assure food security for all people in coming years. The action required is not new or unknown. The action program outlined earlier will require all relevant parties—individuals, households, farmers, local communities, the private sector, civil society, national governments, and the international community —to work together in new or strengthened partnerships; it will require a change in behavior, priorities, and policies; and it will require strengthened cooperation between developing and industrialized countries and among developing countries. The world's natural resources are capable of supporting sustainable food security for all people, if current rates of degradation are reduced and replaced by appropriate technological change and sustainable use of natural resources.

We have the means to assure a food-secure world; let us act to make it a reality for each and every person.

SOCIO-PSYCHOLOGICAL ASPECTS OF INFORMATION IN A DEMOCRACY

by

Edward Wenk, Jr.

INTRODUCTION

We celebrate our times as an information age and interpret the treadmill of new machines as a harbinger of progress. Such rapid change has been exuberantly adopted as though modernism is always better. Like all technology, however, the information age has its own surprises, its unintended consequences. These are still unfolding. Some are ominous.

Given the euphoria seeded by what information technology (IT) does FOR people, it is important to investigate what it does TO people. There is a wide consensus that IT has reshaped and restructured every aspect of personal and social affairs, our thoughts and our way of thinking, our behavior, our work and our fun. It has changed everyone's life, their daily patterns. Because of the speed of electric technology and its global reach across political boundaries, we now experience the "one world" that was the dream of idealists. Information technology, incidentally, is taken here to include all print and electronic communications, for all purposes and by all techniques.

In reality, information technology is the mother of all technologies because its webs function as society's nervous system to inform, animate and synchronize components of life support systems so vital to modern society. IT helps manage national defense, food production, water supply and sanitation, transportation, health care delivery, housing, education, banking, and entertainment.

In an information universe so huge, the scope of this essay has been narrowed to intangible, socio-psychological effects, especially of unintended consequences in the political, social and economic practices of capitalist democracy. That theme connects two immense and complex arenas, of information technology and of a system of government.

IT represents the largest economic domain in America, now approaching one thousand billion dollars annually. It is the fastest growing sector and a conspicuous generator of wealth. IT is also the nervous system of democracy.

Edward Wenk, Jr. *has had a career as presidential and congressional advisor, naval engineer, educator, and author. He is currently emeritus professor of engineering and public affairs at the University of Washington, Seattle.*

The second arena concerns achieving the goals of democracy through the mechanism of government. Little recognized are government's multiple roles to provide national security, safety of life, property and the environment, stimulants to a vibrant economy, to serve as steward of common property resources, as sponsor of leading edge research and of the nation's physical infrastructure, including Internet.

IT is essential to government because democracy cannot function without information, and because information is government's primary product. Indeed, the government's agenda is already loaded with IT-related policy issues such as licensing of channels for high definition TV, suits against Microsoft for monopolistic practices, protection against illegal wire tapping, cable and phone overcharging and invasion of privacy. As the operating arm of democracy, government is mandated to preserve core values of liberty and justice to which a diverse society subscribes. Unintended consequences of IT must therefore be examined to separate those which conspicuously strengthen democracy as a source of citizen empowerment against those more subtle features that inadvertently weaken them.

For information technology to threaten democracy is a paradox. A nation of, by and for the people, is based on a principle that the governed act with the consent of the governed. That depends critically on the body politic being fully, promptly and truthfully informed. With radio, then television including special channels such as C-Span and CNN, and now Internet, that capacity to inform has exploded. At stake in the modern lexicon, however, is whether the notion of "consent" of the governed embraces "informed consent." To that end, technological leverage to generate and disseminate information would seem unequivocal in benefits. That achievement depends on two things, however, on the credibility of information available, and on a prepared mind to process it to knowledge, then to understanding and to action. Part of this dynamic is intellectual, but much is emotional.

In short, every person and every institution has experienced a massive paradigm shift in what was termed four decades ago as the "Information Age." Not clear are destructive influences of information technology that pack a psychological wallop.

The literature on this issue is unfortunately thin. Consequently, this survey is more empirical than scholarly, drawing where possible on every day experiences that involve less the technical understanding of information science and engineering and more of human nature.

A TAXONOMY OF INFLUENCES OF INFORMATION

The transmission of information by electronic techniques now occurs at the speed of light that (1) jars a normal cadence of human activity beyond natural rhythms; (2) squirts a firehose of undigested information overwhelming the human capacity for processing; (3)

compresses time available to test the truth of raw information and to transform it to knowledge and then to understanding; (4) coerces responses to messages so swiftly that decision making shortcuts critical analysis of options and consequences; and (5) generates hectic information environments for both work and play that increase stress, degrade individual performance and undermine personal satisfaction.

In short, the speed of transmission is so enormously increased by technology that human perceptions of time in the industrialized world have been significantly warped.

So have perceptions of space. The increased size, distribution and diversity of information banks and networks can lead to (1) a babel of incoherence and frustration that triggers unnecessary conflict; (2) increased complexity that degrades clarity of message content; (3) vulnerability of systems to malfunctioning from human and organizational error; (4) liability of sabotage and system breakdowns through chaos effects of minute disturbances; (5) misunderstandings from diversity in cultures and languages of participants.

Invisible webs in cyberspace create a new world without borders, economically if not politically. Between the time warp and space warp, we have a world of telecommunications where events anywhere in the world are known everywhere, and instantly.

Moving on, the relative anonymity of message origins opens: (1) doubt about the identity of sources and their credibility; (2) opportunities for malevolent abuse through viruses and break-ins; (3) deliberate anti-social misinformation or propaganda; (4) a communications mode of "road rage;" and (5) difficulty in protecting intellectual property.

Consider also with TV and graphics that the medium can totally obscure the message. This condition: (1) admits the psychological power of images over words to exaggerate, subvert, suppress or block the message; (2) alters the information processing capacity of recipients so as to incapacitate critical thinking, increase vulnerability to propaganda and sales pitches that skillfully exploit emotional responses; (3) teases recipients to seek information packaged as pleasurable entertainment rather than enlightenment; (4) shortens attention spans so as to discount serious content; (5) reinforces insensitivity to socially unacceptable practices of violence or biases based on race, gender or sexual preferences; and (6) leads to depersonalization with a loss in intimacy and in awareness that people are organisms and thus more than information processing machines.

Meanwhile, machines have themselves become emblems of social status, wealth, accomplishment, even of self-esteem. But then people can be safely uninvolved, disconnecting at will.

In a different mode of influence, the zealous merchandising of IT leads to: (1) undue fascination with techniques at the expense of content; (2) pressures to substitute machines and machine-related techniques for human operators. leading to loss in salubrious work environments and in employment security; (3) false feelings of control amidst feelings of loneliness; (4) high anxiety over obsolescence with compulsions to update hard and software and human skills; (5) saturation of Internet with junk e-mail and ads; (6) replacement of ethics of journalism with those of business and commerce.

Not to be overlooked is a reality that large populations on the planet and even in the United States (1) have limited access to modern IT equipment; (2) are limited by education, age, culture and temperament so as to be unable to cope; (3) are then limited in fulfilling their personal goals, achieving career success or in earning capacity, further stratifying populations by wealth.

Finally, just as with knowledge, information technology is about power, economic and political. As a result appetites are tickled (1) to enhance economic advantage through innovation; (2) to enhance economic power through corporate mergers and buyouts so as to reduce competition and gain central control over content and markets; (3) to create trans-national corporations free to dissolve political boundaries in commerce so that business transactions are no longer circumscribed by national creeds and laws; and can move money as information, instantly around the globe; (4) to interlace economic with political power as between corporate and governmental entities where both parties seek advantage in meeting parochial objectives. For the private sector, the goals are higher profit by less regulation and taxes. For government officials, a modus vivendi reinforces public relations by elected officials so as to persuade citizens of the virtue of their past or future actions.

SOCIO-PSYCHOLOGICAL FACTORS THAT BEAR ON THE FUTURE OF DEMOCRACY

Elements in this compendium were intentionally worded to be value neutral, but rapid growth of IT with its swift penetration throughout Western culture demonstrates that whatever the down side, these socio-psychological factors have not dampened uncritical optimism. All, however, have subtle implications for the future of democracy.

The next step, therefor, is to winnow out a few of the most significant influences on the future health of democracy, and to investigate those inimical to that objective and requiring counter-measures.

Vastly simplified, we recognize a strong and overarching psychological link between people and information. People are curious. Instinctively, they want to know. They take pleasure in comprehending the world around them, their prospects for the future. People

like to play, and this desire can be satisfied with electronic IT games. Connecting with other people is enticing, partly from an innate recognition that we are all part of a larger community. But some of this addiction to IT has been learned. There is now a compulsion to stay connected with an electronic umbilical cord using cell phones, beepers, laptop and voice mail and the old-fashioned telephone. Let one of these linkages fail and people panic. That learned behavior is driven by commercialism, the drive for profit by vendors cleverly catering to consumerism.

A key question is which factors actually change people. Because all communications have two elements, of transmission and of reception, as the saying goes, "it isn't what people say; it is what they hear." Altered on the reception side of that couplet is human consciousness of social processes, premises and patterns of thought, and belief structures that control their lives and their perceptions such as of time in human affairs. These patterns of thought are dominated by intense images that command the present without historic foundations, a framework of existing context, or of projections about the future. Today's "breaking news" on television showcases the "now.". Slipping away are the myths and legends by which each generation in the past kept its progeny from reinventing the wheel. Otherwise, we suffer the penalty of repeating blunders at every scale, beginning at the pinnacle of domestic and foreign policy. Cultural survival is drowned out by an impatient and noisy IT revolution. Lost is a dedication to liberty and justice such that high school students in Washington State are suing for exemption from reciting the Pledge of Allegiance.

On the transmission side of the IT couplet, programming succumbs to commercial pressures. Then we face a simplification and corruption of content to suit a television audience of limited literacy, the filtering out of nouns of more than one syllable. Transmissions cater to short attention spans, and lack of critical thinking. Dumbing down has a twin, numbing down. On Internet, messages revert to telegrams, without style and seldom worthy of memorializing. The "delete" function becomes essential.

Analysis of content also reveals a shift in purpose. Once, information was the commodity of learning. Today it is primarily used to market consumer products, to persuade with political spinning of rhetoric, to spread gossip and to entertain, including sports and celebrity glorification. Little contributes to understanding of the world around us, other than by sensational events, and these are seldom treated in depth.

As we explore this link between information and democracy, we are surprised by a paradox. Our society experiences an explosion in communications, but with a pronounced loss in a sense of community.

A SPECIAL CASE OF TELEVISION

Of all the techniques of communicating, the most conspicuous by far is television. More people spend more hours each day engaged in passive pleasures of the tube than with all other media and techniques combined. The average household has a set switched on for seven hours each day. Of particular interest are features of that practice relevant to this inquiry about democracy.

The connection to politics is obvious. Beginning with the 1960 televised debate between Richard M. Nixon and John F. Kennedy, television blossomed as the most powerful medium for candidates to inform the electorate and to solicit votes. These images triggered emotional responses regarding personality that were as powerful as issues and positions. Effectiveness of the medium was confirmed by such techniques as polling and focus groups, creating a cottage industry in its art of persuasion. Then that condition prompted candidates to invest prodigious energies and time in raising funds to purchase time on television, a practice that carries its own seeds of subversion. Funds raised from individual and corporate donors affords them promise of access to candidates and opportunities to influence political thoughts and positions. Campaign contributions now top $200 million per election, yet attempts to corral this practice have, until now, failed.

There are other problems in the political coupling of television and democracy. The most egregious is the inability of viewers to test the truth of what they see or hear. Unlike the special switch parents attach to TV sets to block pornography for children, there is no switch to test the accuracy of the streaming rhetoric. Candidates select photo opportunities and discharge sound bites in the quest for votes, recognizing the demise of literacy where in America twenty percent cannot balance a check book.

These electronic communications reveal another property of the medium. All technologies serve as amplifiers of human capabilities. Television permits national candidates to reach an audience of tens of millions with their theatrics, often in phase with other televised events that polarize thought of a viewer, especially the sensational.

The preoccupation of viewers with the impeachment proceedings of President Clinton confirmed all of these qualities about the hypnotic power of the medium and its exploitation by all the key players.

INFORMATION FOR PROBLEM SOLVING

The most trenchant use of information is for survival. I have the view that the standard definition for survival deserves extension to being both alive and free. Thus, what follows concerns threats to liberty as well as to life.

Detection of danger depends on information, to heighten awareness, to cater a threat analysis, to consider options for action to reduce a threat or to minimize harm if the threat could not be avoided. These all imply short term threat and response behaviors, but the same process applies to longer term predicaments where early warning is sought by sentries, spies or foresight. This simple survival model teaches two lessons. One has to do with information content about danger, but the other is the need for information about the context, that is about the outside world, events that trigger an emergency, the socio-economic setting, its subcultures and the individual movers and shakers in power. Finally in this calculus must be the values that influence how a society thinks and about what, including how it perceives itself in terms of goals and means for achievement .

All of which suggests that in a modern complex world, decision making requires an encyclopedia of factual information, comprehensive, timely and sound. This is what we may know, but there is more. Prudent choice also depends on understanding beyond just knowledge, and that requirement exposes another paradox in the modern world where the gush of information has not been matched by comprehension.

To explain, the first step in information processing is its being organized, refined, purged of anomalies, tested for validity and given a label for retrieval. All of those steps can be subject to operating protocols. The next stage is highly individualistic in converting information to knowledge whereby the previous mechanical steps are further refined and integrated into a person's mind and memory suitable for human reasoning.. Here, knowing assumes a prerequisite of learning, either from education and from experience. Knowledge partakes of the identity, limits to resources and biases of the user.

The next step of understanding is beyond simply knowing so as to interpret the significance of the message in the context of the particular question being addressed, either for survival or for less strenuous challenges. Critical judgment now plays a significant role in testing truth of the input information and connecting bits and pieces to reveal patterns or establish causality. Then follows another step of enlightenment, an epiphany that stems from structures of the mind but also on less tangible properties of intuition and emotion. Even spiritual dimensions of the human experience come into play in the form of moral vision. Imagination plays a key role especially in triggering foresight. This is the basis for action.

Success thus depends on two preconditions, reliable information and understanding. As to the first condition, evidence abounds that truth is not only less respected in our society but increasingly difficult to confirm. Partly this results from the complexity of subject content in specialized fields that defeat general comprehension. But then there is deliberate fabrication. Leaders in all institutional settings when sufficiently stressed seem predisposed to lie. Because this condition is now so widespread, much of the cynicism evident in modern society can be explained. Democracy is threatened if people believe everything they hear, or nothing they hear.

Democracy can survive only by honoring truth. This requirement is based on a foundation of ethics, but it goes beyond. Problem solving in a democracy requires a consensus, usually among stakeholders that differ markedly in their goals, their values, their resources, and their commitment. Only by open communication and civil discourse can progress be achieved. To prove the point, it was uncivil discourse that conspicuously marked the 104th and 105th Congresses.

Interactive communications in problem solving have become ever more necessary in a democratic technological world. The reason is that all technologies have unintended consequences, risks that may be lethal. Most involve technical complexity beyond understanding by the laity, thus requiring the use of expert knowledge. This has always been true in medicine, and the elements of trust has been implicit in physician-patient relationships. Now that is true when considering life safety in planes and trains, environmental safety with oil tankers, safety of nuclear reactors and storage of waste, tendencies toward global warming, the fiscal integrity of on-line banking and investing.

In short, in an intense information environment, in a democracy, far more emphasis is deserved to assure the quality of information rather than quantity.

THE PROBLEM OF MEDIA, AS LENSES, MIRRORS OR FILTERS

Of all sources of information, the public depends for truth primarily on mass media. So do public and private institutions for their functioning. Just consider how leaders at their breakfast reach first for either editorial pages or the stock market quotes. Although businesses hum with management information inside dedicated channels, there is intense use of media products as an intelligence function about competition, uncertainty in public affairs, looming tax policies, and every mode of economic indicator.

Similar information is sought be elected leaders in editorial feedback on their own performance, but also in reporting on salient events, public moods and social as well as economic indicators. Most to the point, all clients need the facts.

The media thus have a seminal responsibility for accurate, objective and even-handed reporting so that citizens can understand issues and options to participate in an oft forgotten reality that American democratic government is self-government. Anchor people and producers in television and managing editors of print media serve as gatekeepers and thus guarantors of fidelity. Because news is now global in scope, the volume vastly exceeds time or space for reporting, they exercise judgment on what people want to know or need to know.

Increasingly, the media are considered to have failed this test with storms of criticism and loss in public confidence. The press concentrates on the sensational, creating a vicious cycle in stimulating appetite for that selection. Televised news is encapsulated as entertainment. This style is intensified by business mergers wherein networks are considered profit making investments. Under pressures to increase audience size and thus attract advertisers, news departments that had traditionally operated with a financial loss were no longer exempt, and were forced to adopt whatever attracted the largest audience. In addition, advertisers have a powerful influence on content and there is concern that news inimical to owners of these businesses is unwelcome.

Investigative reporting had a dramatic success dealing with Watergate so that every reporter seeks to repeat that path to personal glory. However, time pressures preempt double checking of sources especially because competition is now so keen with some sources running 24 hours a day and worldwide. Being first may extract a toll of integrity that for some reporters has been a temptation to plagiarize or invent stories.

Media serve as lenses but also as filters. The enormous prize of modern communications that should help inform citizens has a down side in distortions of the news. Even conscientious citizens are challenged as to wherein lies truth. The ethics of journalism have been replaced by ethics of the market place. In business, this is represented as the drive for profits or enhanced recognition on Wall Street, achieved by skillfully stimulated consumerism. At election time, candidates imitate the advertising of soap products in trying to persuade. Information is crafted, not for edification but for uncritical knee-jerk response.

All of which dilutes content to match the least thoughtful client, to compress messages and then allow the shortest possible time for responses without cogitation. The staccato of images and words relieve the viewers of thought, and in politics relieves the candidates of the stress to sort out issues of greatest portent and explain positions adopted.

The new capabilities had a precedent when both coasts were first connected by telegraph a century and a half ago. As the story goes, when Oliver Wendell Holmes remarked to Thoreau what a glorious achievement it was for people at opposite ends of the country to be able to talk to each other, the rejoinder was, "What if they had nothing to say."

THE FEEDSTOCK OF SELF GOVERNMENT

Thomas Jefferson argued that without timely and reliable information, democracy would be imperiled. That threat is more critical in an age when more powerful technology amplifies risks to life, to liberty and to the pursuit of happiness, and confers disproportionate power on its caterers.

All technologies spin side effects and these externalities as they are labeled by economists are seldom considered in enterprises driven by profit. Many of these side effects are harmful. The public expects protection. It has learned to depend on government to assess risks to life , options for their containment, and a fair allocation of costs. In a democracy, everyone in the system needs information...

Less palpable risks such as to the pursuit of happiness are more subtle. Mostly they appear as slow crises, as for example putting children at risk from tobacco and other harmful addictive substances. These problems require social policies for housing, nutrition, health care, aging, welfare, job retraining, potentially explosive sources of controversy.

Preserving liberty is the ultimate challenge. Not that we are totally defenseless. Such organizations as Common Cause, People for the American Way, the American Civil Liberties Union try ardently to engage those who would muffle the liberty bell. With a different polarity of values, we find numerous patriotic organizations asserting their role to preserve freedom, and their reach is enormous.

Here we crunch into another enigma. Many of these bodies ostensibly protecting freedom also have an agenda to demonize, weaken and downsize government on grounds that government is the enemy. Nevertheless, being attentive to the numinous mission of the founding fathers, whatever the motivation, is important.

Acceptable levels of risk to any of these sectors of life, pursuit of happiness and liberty cannot be defined by exact sciences. Acceptable risk, what we call safety, is a social judgment and there is no moral equivalent to Newton's laws of motion. If the public is to decide how safe is safe, it needs information on which to make a judgment. That process should also follow the trend in medical practice of those exposed to risk be given the opportunity for informed consent. That is also the principle behind legally mandated environmental impact analysis. Altogether, these circumstances make urgent the availability of facts and trustworthy interpretations to the public.

The public, however, cannot be expected to serve as their own experts to comprehend technology-laden issues and the options. They must ultimately trust others, the experts and the government itself. Disclosures on ethical lapses in government, and in business undermine trust, and that condition underpins survival of democracy itself.

REDEFINING THE INFORMATION SUPERHIGHWAY AS SOCIAL PROCESS

Earlier, a taxonomy was developed of psychological elements of human experience that are impacted by information technology. At a higher level of abstraction, it is important to view IT as with all technologies as far more than hardware, software and technique. Information technology is a social process, driving American culture as the engine of change amidst a swarm of unintended consequences and an amplification of all human proclivities, both benign and dark. Information technology partakes of that condition that includes threats to the health of democracy. The nation's status as the only superpower measured in arms and wealth results from a remarkable confluence of democracy and technology. One provided the ends; the other the means. That, however, is too simplified a description. Progress depended on leadership and on information, especially that of the media. But the encounter of President Clinton with Monica Lewinsky found the press playing to the least common denominator with prurient interest in body fluids and DNA. Its feeding frenzy with interviews and unrestrained speculations made the press a willing partner in this national tragedy. It would seem that the medium of television not only morphed the messages; it morphed the messengers.

With a complex, interdependent world in ferment, it is difficult to make sense of our situation as a democracy. But sense we must seek, because we will never return to a simpler society. Amidst confusion, one thing should be clear. As a nation, our most precious possession is freedom, the prize of our democracy. We need, however, renewal of our shared vision, where information for civic literacy and civic discourse play crucial roles.

Psychological aspects of information technology as portrayed earlier represent keys to that renewal. This means facing up to information overload and its pathologies; distortion of facts by purveyors having conflicts of interest; the role adopted by the media as fight promoters, tickling the electorate as entertainment that inadvertently feeds cynicism.

The dominant issue concerns loss in truth and in trust. With citizens paying less attention to their transformation by information, trends will continue of a nation of know-nothings, with arrogant technical wizards focused on the machinery to make investors happy about their portfolio. That condition cedes power to the few who control the nation's information resources. It sets the stage for Orwell's fictional projection of slavery in 1984 through control of information.

We have been reminded that knowledge is power. That concept translates to the notion that whoever controls technology controls the future. They will command all the levers and the tools, and can be expected to function with commercial imperatives that engrave the profit making sector. One can only hope that these leaders of humankind's destiny do not choose a collective future based on a cost/benefit analysis.

With the overwhelming focus on the transmission side of communications, on the hard and soft ware of information technology, we now must focus attention on the human elements in reception. The challenge is how to synthesize 21st century hard and software with human nature that dates back forty millennia. Fortunately, the human psyche contains Divine gifts of ingenuity, free will, hope and love. These spiritual qualities need to be recruited with the intellectual to generate civic literacy and activity in the civic public square.

Once threats to liberty and justice are recognized, there must follow that remedies do not lie in more technology. The problem is us. So is the solution.

REFERENCES

Barglow, R. *The Crisis Of The Self in The Age of Information*. New York: Routledge, 1994.

Deutsch, K. W. *The Nerves of Government: Models of Political Communication and Control*. London: Free Press of Glencoe, 1966.

Dyson, E. *Release 2.0-A Design for Living in the Digital Age*. New York: Broadway Books. 1997.

Grossman, L. K. *The Electronic Republic: Reshaping Democracy in the Information Age*. New York: Penguin Books, 1996.

Herman, E. E., and N. Chomsky. *Manufacturing Consent: the Political Economy of the Mass Media*. New York: Pantheon Books, 1988.

Jones, S.G., Ed. *Cybersociety: Computer Mediated Communications and Community*. Thousand Oaks, Calif.: Sage Publicxations, 1995.

Kurtz, H. *Media Circus: The Trouble with America's Newspapers*. New York: Times Books, 1993.

McLuhan, M., and Fiore, Q. *The Medium is the Message*. New York: Random House, 1967.

Orwell, G. *1984: A Novel*. New York: New American Library, 1961.

Shenk, D. *Data Smog: Surviving the Information Glut*. New York: Harper Edge, 1997.

Stoll, C. *Silicon Snake Oil: Second Thoughts on the Information Highway*. New York: Doubleday, 1995.

Trow, G. W. S. *Within the Context of no Context*. Boston: Little, Brown. 1981.

Weaver, P. H. *News and the Culture of Lying*. New York: Free Press, 1994.

Wenk, E. *Margins for Survival: Overcoming Political Limits in Steering Technology*. Oxford: Pergamon Press, 1979.

Wenk, E. *The Double Helix: Technology and Democracy in the American Future*. Stamford, CT: Ablex Publishing Corp, 1999.

UTOPIAS, FUTURES, AND H.G. WELLS'S
OPEN CONSPIRACY

by

W. Warren Wagar

In the two volumes of *Foundations of Futures Studies* (1996-1997), Wendell Bell explores the methods by which futurists can acquire conjectural knowledge, first, of the likeliest futures of humankind; and, second, of the best futures for humankind. Appropriately, he devotes a whole volume to each of these two formidable tasks. As Bertrand de Jouvenel observed many years ago, our principal motivation for studying the future is to help make the future conform to our desires and preferences. Which of all the many credible futures of humankind will give us the world we wish for? And how can we make our wish come true?

It is also appropriate that Bell prefaces his second volume, on normative futures inquiry, with a chapter examining the utopian tradition in ancient and modern literature. Utopias are imaginative architectural plans for the Good Society. They may be pre-scientific, in Bell's sense, but they encapsulate humanity's highest aspirations through the centuries.

I would only add that utopias come in two quite different models, the first found chiefly in the period before about 1775, the second most prevalent during the past two and a quarter centuries. The traditional pre-modern utopian vision, in both Eastern and Western thought, is the *static* utopia, the utopia located in the distant past or in a distant place, the utopia that serves as a timeless standard against which to measure the performance of humankind here and now. The second, a product of modern historicism and the belief in progress and perfectibility, is the *dynamic* utopia, the utopia that may or will exist in future time after certain preconditions have been met and certain courses of action have been followed. The bridge between the two is perhaps the Judeo-Christian vision of a future great felicity as the culminating event in a series of happenings along a rectilinear time-line, from Genesis to Rapture. From one perspective at least, modern dynamic utopias are nothing more than secular mutations of the Heavenly City.

It follows that one of the most distinctive features of modern utopianism is the attention it directs to the process of getting from here to there, from a perilous present to a fortunate future. Modern utopias are descriptions not only of ideal worlds but also of the making of ideal worlds, the future-historical story of just how humanity can or will get from here to there. Whether by the baron

W. Warren Wagar, *historian and author, is distinguished teaching professor in the Department of History at the State University of New York, Binghamton.*

Turgot or the marquis de Condorcet, by Auguste Comte or Karl Marx, by Edward Bellamy or William Morris, by B.F. Skinner or Ernest Callenbach, by Arthur C. Clarke or Doris Lessing, the typical modern utopia is vitally concerned with process. And the same is true of our major dystopian visions—from Yevgeny Zamyatin to George Orwell. Most modern utopias and dystopias embody histories of the future, as well as visits to a fixed point in future time.

In other words, the typical modern dynamic utopia is an exercise in normative futures inquiry. One of the many opportunities for interdisciplinary cross-fertilization that most futurists have so far missed is an exploration of the common ground shared by futures studies and utopian studies. Whatever the differences between the two, the intersections are of much greater moment. Let us hope that the first chapter of Bell's second volume is a harbinger of more serious attention by futurists to the utopian tradition.

Attention to the utopian tradition was certainly not lacking in the work of the man I have long regarded as the founder of modern futures studies, the English novelist and public intellectual, H.G. Wells. His *Anticipations of the Reaction of Mechanical and Scientific Progress upon Human Life and Thought* (1902) surveyed the whole 20th Century with remarkable acuity and foresight. His essay *The Discovery of the Future* (1902) called upon scholars to develop a new social science of futures inquiry. In a dozen volumes of journalism and amateur sociology, such as *The Future in America* (1906), *New Worlds for Old* (1908), *What Is Coming?* 1916), *A Year of Prophesying* (1924), and *The Outlook for Homo Sapiens* (1942), he peered tirelessly into the human future and beat the drum for his vision of a holistic world civilization.

But he was also an heir of the utopian tradition in literature, contributing three major utopian novels to the genre: *A Modern Utopia* (1905), *Men Like Gods* (1923), and *The Shape of Things To Come* (1933), as well as several brilliantly innovative dystopian novels, including *The Time Machine* (1895) and *When the Sleeper Wakes* (1899). In Wells's prodigious output—well over a hundred books in all—fiction and non-fiction alternated regularly, almost every year of his adult life yielding one or more volumes of each. For most serious critics, this versatility has been seen as a great fault. They find the majority of his novels too tendentious and his non-fictional works too speculative. The fact remains that no one else in the first half of our century made such a sustained, wide-ranging effort both to probe the future and to define in imaginative detail what the future should be.

In many ways the work that most completely integrates Wells's utopian impulse with his futurism is his largely forgotten manifesto, *The Open Conspiracy: Blue Prints for a World Revolution*. This little book, first published in 1928 and re-issued in various revised editions under various titles down to 1935, is not quite a utopia—certainly not in the same sense as Wells's own *A Modern Utopia* or *Men Like Gods*—not because it fails to contain a vision of the Good Society, but because it focuses throughout on the process of how we can arrive there. Nevertheless, since the dynamic utopia nearly always incorporates a narrative that takes us from here to there, such narratives clearly fall within the purview of both the student of utopia and the historian of normative futures inquiry.

The importance that Wells attached to *The Open Conspiracy* may be judged by the number of editions it went through. He was by temperament a rapid, impatient writer who seldom revisited his texts. Once he wrote something, he published it, and then promptly moved on to the next project. Lingering and dawdling and fussing over a book was not his way.

But *The Open Conspiracy* proved to be an exception. In 1930, two years after bringing out the first edition, Wells published a second edition with the revealing additional subtitle, *A Second Version of This Faith of a Modern Man Made More Explicit and Plain*. This also did not satisfy him, and in 1931 he produced an enlarged, revised third edition, now re-titled *What Are We To Do* with Our Lives?, which in 1935 reappeared as the 55th volume in a popular British series, *The Thinkers's Library*.

What we have in all three editions of this singular book is a narrative in non-fictional form of the making of a utopian future. As I shall argue later on, *The Open Conspiracy* is, in my judgment, much more than a historical curiosity or a quaint forerunner of today's futures studies industry. It is a political and cultural manifesto of urgent relevance to our species as we stand on the threshold of the 21st Century.

In brief, what did Wells have to say? He opened his third edition, the one I shall cite here, with a succinct overview of "The Present Crisis in Human Affairs." The year was 1931. Until late September, 1931, when Japan began its invasion of the Chinese province of Manchuria, the world was more or less at peace. The apocalyptic scale of the Great Depression, which had just started, was not yet apparent to anyone. In retrospect Japan's unpunished aggression marks the beginning of the grim "dégringolade" of the precarious post-Versailles world order, but no one living in 1931 could have known that.

The crisis to which Wells referred was a crisis not in this or that place, but in the whole structure of the international system. From 1914 onward, Wells had been among the most vocal supporters of the idea of a postwar League of Nations, but he was also one of the first to brand the actual League that opened shop in Geneva in 1920 a pitiful travesty of world governance, hopelessly inadequate to the tasks at hand. In Wells' view the continued division of the world into a swarm of armed sovereign national states beholden to no higher authority guaranteed future war, just as it had caused the Great War of 1914-1918. The management of these nations by an elite of politicians in the service of nationalism, capitalism, and entrenched privilege guaranteed future social injustice worldwide, just as it had perpetuated mass misery throughout the 19th Century.

Wells's remedy was blunt. He invited men and women of vision, intelligence, and expertise to join a worldwide "Open Conspiracy" to overthrow the existing world system. The "functional" classes, as Wells thought of them, the people who possessed the actual know-how to run the world's business—scientists, engineers, doctors, managers, inventors, builders, and the like—were enjoined to scheme together to bypass, if possible, and displace, if necessary, the old elites. One is reminded of the similar plans of an earlier utopian socialist, the comte de Saint-Simon, and his division of humanity into two classes, "les industriels" and "oisifs," the busy people who actually do the world's work and the indolent people at the top of the social pyramid who claim most of its wealth.

In the chapters that ensued, Wells envisaged his Open Conspiracy as a loose-knit organization operating in the full light of day. It would conspire, but conspire openly, through the media, through the schools, in all available venues, and across all national boundaries. It would begin as a spiritual and intellectual reawakening, a reinvigoration of the secular world-view of modern science. In the vision of science all men and women everywhere were brothers and sisters, members of a single species with the capacity to transform their planet into a glorious garden with freedom and abundance for all. Such a vision was for Wells not merely attractive: it was his faith, his religion, the successor of all the dreams of all the religions of humankind.

But of course the faith of the Open Conspiracy was not an end in itself. It meant nothing if it failed to stir conspirators to concerted action on a global scale as soon as possible. The goal of such action was the establishment of what Wells termed "a scientific world commonweal," not a democratically elected world government analogous to the existing governments of the United States, Great Britain, or France, but rather the global management of human affairs by "suitably equipped groups of the most interested, intelligent, and devoted people ... subjected to a free, open, watchful criticism." In short, by degrees and in due course, the Open Conspiracy itself would become the central guidance system of the world, pledged to fair, efficient administration of public life. Existing national governments, and their ludicrous puppet, the League of Nations, would be dissolved. The old armed sovereign states would vanish. Did Wells imagine that all this could happen with a swish of his wand? Not at all. He devoted several chapters to the obstacles that blocked the path of the Open Conspiracy. In what he called the "Atlantic nations," where he expected the Open Conspiracy to begin and to make the most rapid progress, the old ruling elites would not go quietly into oblivion. Wells entertained high hopes for the more enlightened and liberal sort of emergent multinational corporations, whose captains, he suggested, might be among the first to see the advantages of scrapping the fragile present-day world order. But the crusty old guard of politicians, diplomats, generals, bishops, landed aristocrats, and conservative businessmen who were committed to maintaining the status quo would resist fiercely.

Outside the Atlantic community, the Open Conspiracy would encounter stout opposition from the teeming nations of Asia, Africa, and Latin America, most of whom (in 1931) were still part of this or that European colonial empire or sphere of influence. Not all the peoples of these less developed parts of the world would be opposed. Progressive elements would see membership in the Open Conspiracy as a way of both escaping the sway of their imperial masters and of transcending the squalor and obsolescence of their own traditional cultures. But many others of a more conservative stripe would fight the Open Conspiracy, either in the service of European masters or in defense of antiquated ways of indigenous life, or both.

So Wells concluded that the Open Conspiracy would not be allowed to go about its educational and organizational work unhindered. Its experts might infiltrate governments and corporations and gradually erode their power, but at times the old order—both in the South and East and in the West—might conceivably rear up and show its fangs. In such cases, the Open Conspiracy had to be well prepared for battle, in the most literal sense of the word. If absolutely necessary, it would need to form armed militias to counter the physical resistance of the old order—or, as he put it, "nationalist brigandage." Just how such militias might spring into existence and who would finance and command them, Wells did not specify, but at least he was not oblivious to the threat posed by national states and their armies.

At the same time, although later in the text he questioned his own optimism, Wells expressed the reasoned hope that the democratic states of the Atlantic world would not oppose the Open Conspiracy with force. Looking at Britain, France, Germany (then the democratic Weimar republic of pre-Hitler days), and the United States, he opined that the weight of progressive common sense would, in the main, prevail against the old order without resort to violence. It was more likely that the Open Conspiracy would have to use force to disarm and pacify oppositional elements in Asia, Africa, and Latin America. In any case, the Open Conspiracy would not subscribe to pacifism. Anticipating Winston Churchill's great 1940 speech to the British Parliament, Wells warned: "The establishment of the world community will surely exact a price—and who can tell what that price may be?—in toil, suffering, and blood."

Now what, you may wonder, is the relevance of *The Open Conspiracy* in the vastly different world of 1999? This is surely a fair question, but I challenge its premise. I submit that our world is not so vastly different. Although Bolshevik Russia is long gone, it remains unclear what will replace it, and at this writing a neo-Bolshevik Russia would seem the likeliest bet. Republican Weimar Germany is now Republican Bonn/Berlin Germany, democratic and capitalist much like its predecessor. The League of Nations is now the United Nations, a somewhat more effective but otherwise comparable collection of diplomats and civil servants. The sovereign nation-states against whom Wells thundered have relinquished some of their power and influence to the great multinational corporations, but it is premature to proclaim, as some social scientists do, that the corporations have supplanted them. The states retain their vast budgets and bureaucracies, their immense bristling armed forces, and the essence of their sovereignty. What Benjamin R. Barber (in his *Jihad vs. McWorld*, 1995) has aptly called "McWorld," a globalized, standardized, mass-consumption culture marketed by the multinationals, spans the planet; yet anyone who looks back at the world of 1931 can see its already well-sprouted seeds. Hollywood, Ford Motor Co., and Coca-Cola would not have been mysterious entities to anyone living in 1931.

If we turn our gaze to Asia, Africa, and the Caribbean, everything seems to have changed, tumultuously. The European and North American empires of yore have been replaced by a glittering panoply of sovereign successor states, more than a hundred in all, each with its own flag, anthem, postage stamps, beauty pageant contestants, and seat in the United Nations. But the imperialism of 1931 survives lustily in the neo-imperialism and McWorldism of 1999. The nations and peoples and technologies of Wells's Atlantic community still pipe most of the tunes and enjoy most of the discretion and initiative as we plunge into the 21st century. I sympathize with Ziauddin Sardar and his colleagues (in their new book, *Rescuing All Our Futures*, 1999), when they seek to change the provincially Western character of futures studies. But they have their work well cut out for them. I also sympathize with Andre Gunder Frank (in his path-breaking book *ReOrient*, 1998), when he maintains that Western hegemony in the world-system is quite recent, a mere blip in the steady pulse of world-historical Asian preeminence. But that "blip" shows no signs of going silent.

My heartfelt conclusion is that the world of 1931 and the world of 1999 are lamentably much the same, that the threat of interstate violence has not disappeared, that imperialism and the armed sovereign state are not dead, and that the "Atlantic (or American) Century" could well extend far into the 2000s. In sum, if Wells's prescription for world revolution made sense in 1931, it should still make sense today.

In the short run, it is arguable that we need more U.N. and N.A.T.O. fire-fighting operations around the globe to maintain the precarious stability of the world-system. If Wells were still alive, he might not have opposed the efforts of the Atlantic nations in the 1990s to "pacify" Panama, Iraq, Somalia, Bosnia, and the like. He would not have mistaken any of these interventions for the Open Conspiracy, but he might have seen them as crude anticipations of the new world order of his prophetic vision.

In the long run, however, Wells would have wanted us to do much better. Make no mistake: I am repelled by his suggestion that the Open Conspiracy should bypass the democratic process. The only excuse I can make for him is that 1931 was a dark year for believers in democracy. Yet it is also certainly true, and has always been true, that revolutions need elites. Spontaneous mass uprisings invariably result in the ruthless repression or slaughter of the masses. Acephalous movements are never effective. Today more than ever, we need an organized, self-aware Open Conspiracy of men and women of vision and courage who can lead humankind to something not unlike the Cosmopolis that Wells envisaged at the end of modern history's trail. The so-called "advanced" Atlantic nations may or may not supply the chief leaders of such a 21st-century Open Conspiracy. It would not trouble me in the least if all or most of such leaders came from Asia, Africa, and Latin America. I suspect it would not have troubled Wells, if he could speak from his resting place. He would have been surprised, but not, I think, troubled.

Of course from our perspective in 1999, Wells's Open Conspiracy is a pipedream, a will-o'-the-wisp without a shred of plausibility or tangibility at our moment in world history. But what are the alternatives? National self-interest and corporate greed cannot provide rational or humane or just solutions to the environmental, social, economic, and political quandaries of the 21st century. Neither can the dark turbulent forces of what Barber has labeled "Jihad" (a word he uses in the generic, not in the literal sense). The nativist cultures that oppose globalization, including Western fundamentalisms, are not powerless; but they do not speak for humankind. They speak for the antediluvian utopias of pre-modern times. They are divisive, mutually antipathetic, and irrational. They represent a dead end. They will not prevail.

I can only hope that as the next century wears on, the gist of the argument of H.G. Wells's masterpiece, *The Open Conspiracy*, will reappear in some form or other, and help to inspire the kind of world revolution he dreamed of. The vision of what I once called the City of Man and would now call the Human Commonwealth— social-democratic, liberal, and secular, led by our best and bravest Open Conspirators—remains humankind's most hopeful chance of surviving into the 22nd century.

THE ABSOLUTE URGENT NEED FOR PROPER EARTH GOVERNMENT

by

Robert Muller

Since globalization is the primary evolutionary phenomenon, challenge and opportunity of our time, it obviously raises the extremely important question of the type, role, structure, strength, and resources of the international system. And since the Earth is in peril and the greatest part of humanity is still in misery, the remedies must be audacious and strong, even if they seem unrealistic or difficult to accept by those in power. We must stretch our minds and hearts to the dimension of the problems. As President Roosevelt wrote in his own hand on the day before his death for the speech he was to deliver at the opening of the San Francisco Conference convened to give birth to the United Nations: "The only limit to our realization of tomorrow will be our doubts of today."

In my view, after fifty years of service in the United Nations system, all the above points to the utmost urgency and absolute necessity for proper Earth government. This should become the priority item on the agenda of world affairs for the year 2000. The poor countries who have been waiting so long for world justice should be the first to request it after 50 years of promises from the rich countries.

There is no shadow of a doubt that the present political and economic systems—if systems they are—are no longer appropriate and will lead to the end of life evolution on this planet. We must therefore absolutely and urgently look for new ways. The less time we lose, the less species and nature will be destroyed. I would urge therefore that consideration be given to the following:

1. A World Conference on Proper Earth Government through the Free Market System

Since business was the first to globalize itself worldwide, far beyond governments, and since corporations are now for all practical purposes ruling the world, we should give them the opportunity, even request them to assess their full responsibility for the future of all humanity, all living species and of the Earth itself and prove to us the validity of their claim that the free market can do it all.

The world corporate community should be asked to answer how it would provide for a well-preserved planet and the well-being of all

Robert Muller *is chancellor, United Nations Peace University, and former UN Assistant Secretary General.*

humanity, full employment, the renewal of natural resources, the long-term evolution of the planet and continuation of life on it, the real democracy of the consumers in a corporate power and wealth economy. Such a conference would bring together the heads of the major world companies, banks and stock-exchanges, the World Bank, the IMF, the GATT, the new World Trade Organization, the International Chamber of Commerce and similar organizations.

2. A Second Generation United Nations for the 21st Century

Since the United Nations is the only worldwide, universal organization at present available, since it had fifty years of valuable experience and many successes, since it paved the way to proper Earth government, instead of putting it on the defensive, unjustified attacks and criticism, reduction of resources and non-payment of obligatory contributions, governments should honestly ask themselves if a better way would not be to consider a second generation United Nations upgraded by a true quantum jump into a proper Earth preserving and human well-being and justice ensuring organization of our planet.

Such a conference would have at its disposal many proposals and ideas for the strengthening of the UN made by various UN bodies, governments, Secretaries General, outside organizations and retired elders like myself. I have formulated many of them in my 2000 ideas and dreams for a better world and will provide samples at the conference.

I recommend the urgent holding of a UN Charter Review Conference and second Bretton Woods Conference to assess the United Nations system's role, potentialities and substantial strengthening to cope with the critical issues and needs of the Earth and of humanity in the future. A Charter Review Conference would moreover give a voice to 134 governments of the present 185 members, which did not participate in the drafting of the Charter and creation of the UN. The UN Secretary General, a member government or a group of governments should request the inscription of this item on the agenda of the next UN General Assembly.

There is no doubt that given the massive changes which have taken place since 1945, an Organization created 53 years ago can simply not be adequate to deal effectively with the mounting, unprecedented and massive world problems of a new century and new evolutionary phase of our planet.

3. A New Philadelphia World Convention for the Creation of the United States of the World

The star-performance, often called "miracle" of the American States in the Constitutional Convention in Philadelphia 200 years ago which put an end to a similar political chaos in North America between numerous, sovereign independent states at the time, should be repeated.

Such a Convention of all nations would review the state of world democracy and would have to add to the system of balance of powers the new dominant power of business. "Philadelphia II" is a project of US Senator Mike Gravel who proposes a convention for the writing of a charter for a Global Constitution.

4. A World Conference of All World Federalist and World Government Associations and Movements, to Propose a Federal Constitution and System for the Earth

An immense work has been accomplished by the World Federalist Association headed by Sir Peter Ustinov, by its national associations and many other world government movements. There exist already several draft World Constitutions. World philanthropists should sponsor a World Conference or other ways to come up with a World Constitution of the 21st century. We may remember that during World War I, Andre Carnegie brought over to the US two Belgian scholars who drafted the statutes of the League of Nations and earned the Nobel Prize for it. Contemporary philanthropists should be inspired by such examples. According to UNESCO only 15% of philanthropy is international and most of its is bilateral. When the global world and the human family are in greatest need, they are the orphans of philanthropy.

And is it so inconceivable that two big federal countries like the United States and Russia might take the initiative of calling a world conference for the establishment of a global federal government in their image? After the Cold War, what a warm spring, a spring of truly united nations this would be for our precious planet!

5. A World Conference of the creation of a World Union on the pattern of the European Union

The world has recently witnessed another political miracle, similar to the American miracle in Philadelphia: The miracle of Strasbourg, the birth of the European Union of 15 European countries which have finally put an end to their antagonisms and wars, decided to unite and cooperate and have abolished the borders between them. Every European can now settle anywhere in the Union, elect a European Parliament at the same time when electing his national parliament, and can have his government condemned by a supranational European Court of Human Rights when his rights are violated. Also, the European Union has its own European budget and tax system and is not dependent on national contributions as is the United Nations. In 1990, the European Economic Community had already a budget of 7.4 billion dollars, ten times the UN budget for all activities. This example is so hopeful, so powerful, so novel and inspiring that I recommend it as an outstanding guidelight for more regional communities and for the entire globe.

It is significant that the European parliament has called for the setting up under the auspices of the United Nations, of an International Environmental Court and a World Environment Agency, of which the European Environment Agency would be a regional branch. It also wants that consideration be given to the setting up of a Parliamentary Consultative Assembly within the UN. We should wholeheartedly support these proposals.

I recommend that the European Union organize meetings and conference with outside countries to show them how they can move towards regional unions and how a World Union can be established. This would render a great service to the world and to the UN General Assembly.

6. A World Conference of the Planet's Five Continents for a Proper Earth Government through Continental Unions and a World Union

About ten years ago or more, I suggested to President Bush that in view of the creation of the European Union, the American countries from Alaska to Tierra del Fuego should create an all-American community or union. He listened to me but instead of creating that community in a common, joint effort of all American countries, as was done in Europe, the US negotiated separate trade agreements first with Canada, then with Mexico, and now the Latin American countries created their own Mercosur (the southern Latin American market) and the future of an American Union is in doubt.

It might be noteworthy that indigenous people of the Americas believe in a prophecy according to which the Eagle and the Condor will meet on sacred Mount Rasur in Costa Rica from which a civilization of peace and nature will spread to the entire world. It is on that hill that the dream of the demilitarization of Costa Rica was born, and where the United Nations created the first University for Peace on this planet, as well as an International Radio for Peace. (The Earth Council created by the Rio de Janeiro conference will also move to it.) Simon Bolivar, for his part, in his dreams, prophecized that someday the capital of the world will be located in Central America.

The continental approach to a world union remains an important avenue. One could conceive five continental unions: the European Union, an American, an African, an Asian, and an Australian Union. A World Union could be constructed as a super-structure and common political system of the five continents.

7. A World Conference of Earth and Human Government through New Bio-political Modes Patterned on Examples from Nature

A very novel approach to the organization of humanity and its proper relations with the Earth and nature is to follow the biological models offered by the formation and admirable functioning of numerous colonies of cells, bacteria, and living species observable in nature and now well studied. This is a very advanced science which opens up the most interesting and promising vistas. A bio-political science can and should now be rapidly developed on its basis. It would offer a very much needed bio-political revolution of the Earth's political system and science. Here the Earth and nature would come to their full preeminence and rights. All other world governmental avenues will sooner or later lead to it. First models are already the bio-regional approaches existing in certain areas of the world such as the Arctic Forum and the big river basins and mountains chains cooperative agreements.

Beyond this bio-regional vision and approach is the idea and proposal of Barbara Gaughen-Muller to create a United Nature, a transformed United Nations to respond to the fundamental unity of nature of which humans are part. Humans would not dominate nature but cooperate with it and learn from it. It is probably the most advanced, timely and imaginative vision of the total, proper functioning of planet Earth.

The Natural Law party created by British scientists, which exists already in 85 countries and has become the third largest party in the United States could be the spearhead of this new approach.

A World Conference on Proper Earth Government through What the World's Religions Have in Common in Terms of Universal, Global Spirituality and Worldwide Human Experience

Last, but not least, humanity has reached a point when we must consider our human presence, past, present and future on this particular planet in the universe. We have now tremendous information on the universe in which we live. In addition to our total consciousness of our Earth and its global evolution we are also now acquiring and developing a cosmic consciousness of the universe. This is one of the greatest advances in human history. But the mysteries of infinity and eternity will probably remain beyond human and scientific grasp. This has the result of bringing together the spirituality or basic "faiths" of all religions and science. God, the gods or the Great Spirit and their emissaries, prophets, and human incarnations like Jesus gave humanity at its early stages a cosmic, universal, all-encompassing faith or feeling for the mysteries of the cosmos, for the norms of love and for the miracle of life and norms of behavior between all humans, other species and nature.

The messages or "revelations" should not be neglected. They contain some of the profoundest answers to human behavior, fulfillment and survival. Great was our astonishment in the environmental crisis to discover the wisdom and rules of behavior towards nature dictated by the Great Spirit to the indigenous people of this planet, and towards Creation in practically all religions. The world's 5000 religions are filled with incredible wisdom regarding human morality, belief in life, environmental adaptations, survival and future evolution. This is strongly coming to the fore at this time in the following:

1. The dream and plan of Robert Schuman, my compatriot from Alsace-Lorraine, to see the European Union, which started with a coal and steel community followed by an economic community, followed by a political union, culminate in an all spiritual European Union including the Eastern European countries, especially "Holy Mother Russia." For him this was more important than the extension to these countries of a military union through NATO.

2. The San Francisco Initiative to create a United Religions Organization similar to the United Nations (also born in that city), where all religions of the world will cooperate, define what they have in common, provide their wisdom on human behavior and morality, and right relations with nature, God's Creation and the universe thus ushering the world into a great Spiritual Renaissance. In the process they will hopefully reduce and progressively give up their fundamentalism in favor of a global spirituality; the same way as nations in the United nations have reduced to some extent their national fundamentalism call sovereignty.

3. In August 1998, at the 20th World Congress on Philosophy in Boston, a World Commission on Global Consciousness and Spirituality was created. Karan Singh of India and I are its Co-Chairmen.

4. The convening in Pretoria, South Africa, in December 1999, at the invitation of Nobel Peace prize winner, Bishop Edmund Tutu, of a third World Parliament of Religions. The first such Parliament was held in 1893 and the second in 1993, both in Chicago.

This global religious cooperation towards a spiritual renaissance is accelerating.

THE NEED FOR A CHANGE IN VALUES AND BASIC RETHINKING OF ALL PRINCIPAL SEGMENTS OF HUMAN LIFE

In recent years, Erika Erdmann, the research aide and librarian of Nobel Prize winner Roger Sperry, and professor Jean-Claude Leonide, a reputed French anthropologist, undertook a survey of long-term evolutionary scientists which showed that scientists were becoming more optimistic as a result of the birth of a global consciousness which makes us humans aware of our mistakes and problems and helps us solve them by changing course and adapting to evolutionary requirements. Their survey revealed that the theory of "chaos" according to which the universe and human life make no sense is losing ground. The new theory is that on any planet having life in the cosmos one species sooner or late evolves to a point of gaining a total knowledge of the planet it lives on. It will then be in its power either to continue evolution or to bring it to an end. The first course will require that the former values of that species, values not respectful of the new phase of evolution, must be replaced by new ones which take that evolution into account. These new values are a major new evolutionary imperative.

In my view and in theirs, humanity has reached that stage on Planet Earth: we must revise our basic values dating from the 19th and early 20th century and acquire a new evolutionary wisdom which respects nature, the Earth and its basic laws. If nature has produced the incredible, sophisticated variety of innumerable living species around us, each one a true miracle, it is simply not possible that the human species is not a miracle too, perhaps the most advance of all. We are no longer our own objective. We have become the caretakers, the trustees, the shapers of future evolution, the instruments of the cosmos, integral parts of it, as we have already recognized of late to be to of the Earth.

The future of Earth will be bright and life will not become extinct if we decide so on the eve of a new century an millennium. We are entering a thrilling, transcending new global, cosmic phase of evolution in the line indicated by Teilhard de Chardin, the anthropologist, if the human species understands its suddenly momentous, incredibly important evolutionary role and responsibility.

NOTE

1. This is an abridgement of Chancellor Muller's "The Absolute, Urgent Need for a Proper World Government" which is presented in full in his recent *Two Thousand Ideas for a Better World, A Countup to the Year 2000*. Vol. IV, Ideas 1901 to 2000, can be ordered from the UN Bookshop, New York. Tel: 1-800-553-3210.

THE NEXT 1000 YEARS: THE "BIG FIVE" ENGINES OF ECONOMIC GROWTH

by

Graham T.T. Molitor

The previous thousand years, and longer, has been dominated by four successive "waves" of economic change that dominated civilization and permeated every facet of society—agriculture, industrial manufacturing, services, and the current knowledge-information-education eras. Over the impending 1000 years, humanity will accommodate at least another five waves of economic dominance: leisure (dominant sphere of activity by 2015); life sciences (by 2100); mega-materials (2200-2300); new atomic age (2100-2500); and new space age (2500-3000). Each of these entrepreneurial sectors has been building momentum for hundreds, even thousands of years. Sudden surprises should not catch forecasters unaware of impending "economic centers of gravity."

THE "BIG FIVE" ENGINES OF ECONOMIC GROWTH

Past Waves of the Economy: Old Four

The much-discussed Third Wave broke long ago. Actually, there were four, not three, successive waves of economic development through which advanced economies progressed. Each era was based on a different economic foundation:

1. Agricultural Age

was premised on wresting sustenance and livelihoods from the land. US jobs peaked in this sector during the 1880s.

2. Industrial Era

concentrated on mass production of fabricated products. US jobs peaked in this sector during the 1920s.

3. Service Era

undertakings involved employing the skills of third

Graham T.T. Molitor *is Vice President and Legal Counsel, World Future Society and President, Public Policy Forecasting, Potomac, Maryland.*

party providers to render specialized expertise the consumer could not perform so well by himself. Jobs peaked in this sector during the mid-1950s.

4. Information Era

technologies rely on intellect and knowledge that educate, entertain, and manage human affairs. US employment in this sector has been dominant since the late-1970s.

Attention currently is riveted on knowledge and education made possible by communication and computer linchpins. Still reaching toward its zenith, this current wave of technology has a relatively few remaining years of dominance—possibly as few as 20 years— prior to being supplanted by another surging wave of economic importance. A mere two decades is a short enough span of time to start seriously thinking about what stands in the offing.

What does come next? Looming changes imposed by this on-slaught of new enterprise will promote as well as destroy jobs and earnings, reshape economies, and affect the entire world. Declining sectors displaced by newer ascending technologies will be hardest hit. Planning focused on where things are headed and assessing how to deal with massive change that accompanies such transitions is essential to minimize dislocations. Survival of entire industries, and future economic growth overall, depends on staying at the forefront of emerging technologies.

Each new wave of economic activity will enjoy a brief predomi-nance, similar to the previous Big Four. Successive waves of economic activities, each in its own turn and time, will become the economic "center of gravity" of the economy. Dominance will mark from the time the particular economic undertaking becomes the modal or largest provider of employment. Soon thereafter, that sector will account for the biggest share of gross domestic product.

IMPENDING WAVES OF ECONOMIC CHANGE: THE BIG FIVE

Exactly what kinds of new economic activity come next? What should planners and policy makers be eyeing? Currently on the horizon are five economic thrusts that will dominate jobs, economic output, and GDP. Each one of these five impending "centers of gravity" have been building for many years. Surprisingly, each of these "economic linchpins" has been developing and gathering momentum for as long as a century! Careful research reveals that the roots of fundamental economic change take hold decades, even centuries before breaking, reaching dominance, and peaking. Each of the impending new economic "centers of gravity" already are well along in their early stages of development and drive toward dominance.

The proliferation of start-up enterprise with cutting-edge ideas, a gleam of hope, and enormous dedication to stick with their dreams provide indications of things to come. Early bellwethers of impending technological change include recent innovations in laboratories and among researchers. More tangible and immediate inklings can be gleaned from public records of inventions and patent applications and grants. Assessing investments that risk-taking venture capitalists make in anticipation of imminent breakthroughs likely to catapult a new wave of technology upon society provide other glimmerings.

The five major technologies soon to engulf advanced nations, reshape entire economies, and drastically alter human life include:

1. **Leisure Time Era (by 2015)**

Hospitality, recreation and entertainment. Leisure time pursuits have been a part of human activity from the very outset. The change about to be fully felt occurs when "free time" dominates total individual lifetime activity.

2. **Life Sciences Era (2100)**

Bio-tech, genetics, cloning. genetic engineering, transgenics, and "pharming," among others. Theoretical underpinnings trace back more than a century. The pace began to accelerate with the human genome project, and it reached a dramatic turning point with the cloning of Dolly.

3. **Mega-materials Era (2200-2300)**

Quantum mechanics, particle physics, nano-technologies, isotopes/allotropes/chirality, superconductors, and microscopic imaging systems constitute the major core technologies. This sector began to take off with the development of plastics, bullet-proof Kevlar, ceramic engineering, high-strength alloys, composites, silicon, super-alloys, high-temperature superconductors, crystallography, cryogenics, semiconductors, time/temperature/pressure variable materials, designer materials.

4. **New Atomic Age (2100-2500)**

Thermonuclear fusion, hydrogen and helium isotopes, and lasers constitute the key technologies upon which almost every energy-dependent activity will depend. Paramounce of these activities looms every-closer as finite fossil fuels—first petroleum, then natural gas, and finally coal—are depleted. This Era reaches its apex a century or more into the future. Roots of coming change, however, trace far back in time. Commencing with theoretical foundations, this early phase came of age with "splitting the atom." Early experiments soon led to atomic fission, followed by development of thermonuclear explosives. Breakthroughs essential to harnessing fusion center on advances in magnetohydrodynamics, laser-induced implosion, and quantum physics.

5. New Space Age (2500-3000)

Astrophysics, cosmology, spacecraft development, exploration, travel, resource gathering fare pivotal activities propelling this stage of development. Beginnings for this sector trace back to gunpowder and rockets developments over 2,000 years ago. World War II rockets and jet aircraft accelerated the pace. Sputnik, spy satellites, manned space missions, extra-planetary probes, and telescopic arrays that pierce outermost limits of the universe are among the activities adding to the conquest of space.

LEISURE TIME ERA: DOMINANT BY 2015

Leisure time businesses will account for 50% of US GNP shortly beyond 2015. This near-term timeframe is well within planning range.

Determining the size of this sector depends upon what is counted. Leisure time entrepreneurial activities in the orbit of this next wave of economic activity include: recreation, hospitality, entertainment, gambling and wagering, travel, tourism, pleasure cruises, adventure seeking, reading, hobbies, sports, exercising, games, computer games (hardware, software), outdoor activities, cultural pursuits, theater, drama, arts, poetry, opera, symphony, disco and bands, night clubs, bars and taverns, country clubs, retreats, bird watching, gardening, moving pictures and theaters, broadcast media (television, radio, citizens band transmitter/receivers, shortwave, etc.), visiting and socializing (with family, friends and neighbors), audio and video recordings (including production, distribution, retailing, sales, rentals, etc.) internet and on-line activities, etc. Calculations, it becomes glaringly evident, depends upon how one tallies up the activities.

"Big Entertainment" conglomerates that include film, television, publishing, music, hotels and theme parks—like GE's NBC, Westing-

house's CBS, Disney's ABC, and Time-Warner-Turner and Viacom—
are among the companies already in the forefront of this new sector.
Indicating where the nation is headed is the little-noted fact that 15
million Americans visited Disneyland in Anaheim during 1996, while
only 10.8 million (excluding 9.3 million business visitors) visited the
nation's capitol. Priorities are changing. Even while mass attention
is focused on information era interests, savvy investors have begun
shifting their bets to leisure-hospitality enterprise.

Leisure time growth

What factors gave rise to the increasing dominance of hospitality,
recreation and leisure time activities? This dramatic transformation
follows an incremental and evolutionary pathway. As society
progressed through each of the previous great eras of economic
activity, leisure time increased. Leisure time, continuing to steadily
increase, very soon will account for over 50% of lifetime activities in
advanced-affluent nations.

Ten thousand years ago when homo sapiens struggled to survive,
subsistence was nearly a fulltime effort and leisure time was virtually
nil. When agriculture began to develop 8,000-10,000 years ago, the
reduction in time spent hunting and gathering freed perhaps 10%
percent of lifetime activity allocable to leisure. Although seldom
thought of in this manner, the 10,000-12,000 years of continuing
improvement in agricultural practices, may rank as the most
technologically advanced activity of all! What all of this meant was
that instead of spending nearly every waking moment foraging and
hunting for food and eking out a meager survival, the prodigious
productive might of agriculture freed non-farm workers to pursue
other matters.

In like manner, division of labor among craftsmen and later on
industrial mass production producing "things" people needed,
enabled others to side-step time-consuming tasks third party experts
could perform more quickly and cheaply. The importance of
organized labor cannot be understated. Nobody, no matter how
hard-working, could single-handedly build a jumbo-jet airplane or an
ocean-going steamship in their backyard. Around 2000-3000 BC, as
craft specialization commenced, time-consuming fabrication of
"things" permitted as much as 17% of a lifetime to be spent on
leisure time activities. Between BC 1 to 1600s AD, machines, mass
production, and automation freed humans from still more drudgery.
As working time waned, more free time—17-22% of a lifetime
—opened up for leisure pursuits. By the 1700s powered-machinery,
including primitive steam engines, got things done quicker, and freed
more time. This enabled as much as 23% of a lifetime to be dedicated to leisure interests. By the 1900s, electrically-powered machines,
at work and in households, further reduced time required for both
chores and work.

Most everybody has to work in order to earn their keep. Next to sleeping, workhours in gainful employment and on the job have made the greatest demand on available time. But the time spent at work also has steadily declined. The end of that trend is not in sight.

Workweeks

Steadily declined from 69.8 hours in 1859 to less than 40 hours by the 1990s. Portents for further reductions are telegraphed by 30 hour workweeks in Europe. The pattern for reduced workweeks typically commences with hazardous, unusually arduous or demanding jobs, and to accommodate working mothers. Beyond this, heightened by pressures of high unemployment, 20 hour workweeks are being proposed and debated, especially by labor-dominated European political parties.

Holidays

Which have steadily become more numerous, also add to leisure time. Federally recognized holidays in the US totalled ten days in 1998. State and local jurisdictions often observe a few more. While Americans enjoy only ten federally-recognized holidays, residents in other nations enjoy up to 18 holidays. This disparity suggests the probability of Americans eventually gaining a few more official federally-recognized holidays. Adding impetus is the impact of diversity and multi-culturalism. In a mosaic society, numerous racial, ethnic and religious interests indicate more holiday celebrations yet to come.

Vacations

Also will grow longer. Americans, currently earning an average of twelve days vacation time, are headed toward 37.5 days. That is the number of vacation days Finnish workers enjoyed during 1993. The Finns are not alone. Tendency toward this goal is bolstered by liberal vacation policies in other Western European nations where two to three times the number of vacation days as American counterparts were enjoyed in 1994:

32.5 in The Netherlands;
32 in Italy;
30 in (West) Germany;
27 in Sweden;
25 in Great Britain, France, Denmark
24.5 in Spain;
24 in Switzerland.

Leaves of absence

With or without pay, also continue to become more generous. Provision for time off is growing. Voting, military, reserve/duty/call-up, jury duty, medical appointments/sickness, funerals and bereavement, personal leave, maternity leave, parental leave, sabbaticals, rest time, lunch time, recreational facility use (in house), and so on, are among the reasons allowing time-off from work.

Retirement

Trends also affects leisure time. The current trend allows retirement at an ever-earlier age. Retirement age between 1950-55 averaged 67 years, dropped to 63 years, 1985-90, and to 60 years (mode) during 1994. Social Security increases eligibility from 65 to 67 years by 2022. Retirement age that has come down, however, will go back up. By 2025 retirement age may rise to 70 years, as a result of longer and more productive lives. Assume, for the moment, that life extension becomes widely available and individuals live to be 160 years old; if that person retired at age 60, how would the last 100 years be spent? Considered from this vantage, the basis for leisure-time pursuits take on a whole new importance.

Trends in other parts of the economy, and in people's personal lives, influence the amount of "spare time" individuals have to devote to leisure pursuits. Decisions to sire fewer or no children, for example, increases a couple's available leisure time. Household chores that used to require the better part of a day to accomplish, could be executed swiftly and effortlessly by a horde of household appliances. Lessened housekeeping workloads, as they are increasingly taken over and performed by appliances or third-party providers (lawn service, maids, nannies, handymen), add further to leisure time possibilities. Increased rate of travel also opens up additional time to devote to other pursuits that otherwise would be spent getting from place to place at slower speeds. Motor cars, aircraft and powered transport systems drastically reduced time required for just getting about. Instead of walking 5 miles to the local village store, a person could get there in minutes by motor vehicle. All told, during the 1900s, over 40% of a lifetime was devoted to leisure. Sometime before 2015, over 50% of a lifetime will be earmarked for leisure.

How Americans spend a lifetime depends upon how activities are tallied. In broad strokes, Americans spend the 75-80 years of their lives: one-third sleeping (8-hours daily); one-third (20-25 years) getting an education; one-third earning or enjoying the fruits of labor. Soon the proportion spent in "free-time" will exceed one-half of all lifetime activities. Attributing the time spent gaining an education (essentially the ticket for determining the kind of work pursued) means that nearly two-thirds of a lifetime is either dedicated to or actually spent at work.

Leisure time dominance will usher in a host of new attitudes, outlooks, and activity preferences. Overall, the new paradigm will emphasize experiences, instead of things. Lifestyles, at least for the near term, are likely to concentrate on instant gratification, me-now, hedonism, self indulgence, living for the moment, narcissism, and self-centeredness. Adventuresome, thrill-seeking pursuits will boom, as individuals test their capabilities to the limit and add excitement to homogenized and hum-drum lifestyles. Seeing, first-hand, other parts of the world will soar. China will become the world's top-ranked tourist destination by 2020. For some persons, leisure interests will be devoted to maximizing self-worth and developing inner-potentials. For most, however, leisure time is unlikely to be spent in achieving self improvement or self-actualization. Others will devote more time to spiritual matters. The age-old search for life's meaning and purpose will take on a new momentum.

LIFE SCIENCES ERA: DOMINANT BY 2100

Life Sciences, steadily gaining momentum, will begin to dominate economic activity by 2100, then predominate well into the following century. Mapping the human genome opens up epoch-setting potentials. Gene mapping provides a blueprint for biological sciences akin to the periodic table of the chemical elements. Once laid out, the ability to comprehend and manipulate combinations becomes clearer. The secret of life itself, one of the most sought after mysteries of all time, is beginning to be revealed and understood. Creation's blueprints for lifeforms—biotechnologies and genetic engineering—open up almost unimaginable opportunities for controlling evolution for all organic lifeforms. Fully developed, these technologies entail the power to change the very form, structure, properties, and durability of living matter. Ability to create new lifeforms by cloning makes duplication of genetic successors possible, and shakes the meaning of life to its very roots. Genetic engineering will gain the know-how to control the evolution of plants and animals, and eventually human beings.

As early as 2020, the power to create or to take life will unleash anew the most divisive moral and ethical dilemmas of all time. Eugenics, humans taking conscious control of their evolution, is certain to become the most controversial center of these debates. Organized religion, along with other critics and crusaders, will exert powerful efforts to stymie and stifle life-altering genetic achievements. Similar questioning has accompanied most every major advance in biological manipulations. As human intervention in creating or ending lifeforms evolved, religious leaders always have done so, each step along the way. Not so many years ago, hybridization of plant life by Luther Burbank was denounced as blasphemous by church leaders. Centuries earlier, human dissection was blocked on grounds it was sacrilegious, cruel, immoral and obscene. Obstacles encumbering advances in bio-technologies, genetics and life sciences will be overcome. Eradicating genetic diseases, extending life expectancy, increasing food production, improving pharmaceuticals, and contributing overall to extraordinary advances in the quality of life are too important to be denied.

On a positive note, elimination of genetic disease and dysfunction is within grasp. Businesses already have developed diagnostic slides and films that can be doped with a DNA sample, and run through a mini-lab to ferret out genetic diseases. Business-card-sized probes with one million micro-wells have been designed to assay indicators of disease in a few hours. Persons with discovered genetic defects face discrimination, invasions of privacy, healthcare rationing, abortion, and euthanasia.

Genetically engineered "designer babies," cost-prohibitive to all but the wealthy at the outset, will create other new forms of discrimination. Surgically removed ovum and sperm can be carefully selected and combined in the laboratory to create a perfect zygote. The conceived specimen then may be implanted in the biological parent, a third party, or a surrogate mother host. Sex will become more recreational than procreational. Birth control, contraceptive technologies, and fertility controls will predominate. "Normal" sexual reproduction will be deemed as foolhardy as foregoing prenatal care is today. As early as 2020, social impacts of creating and controlling the beginning as well as the end of life will unleash divisive debate. Threats of genocide, creating a "master race," and the gargantuan geriatric problems brought about by increasing life expectancy to 125-160 years pose a different set of realities. How, exactly, will individuals sustain themselves over a retirement spanning up to 100 years? On another plane, decisions concerning euthanasia are inescapable. Dr. Kervorkian's legal travails, and policy reversals in Oregon and Australia merely test the limits of such policies.

Genetic technologies applied to crop and animal husbandry will far surpass previous yield increases made possible by the green revolution. Bio-engineered crops will be designed to thrive in hostile environments, survive without irrigation, increase nutrient composition, minimize fertilizer and agri-chemical needs, and so on. Genetically modified tomatoes have been developed to resist spoilage and undergo rough handling. Growth hormone therapy has been successfully used to boost milk output of cows. Human growth hormone treatments have been used to overcome "dwarfism." Genetically engineered "pharm-foods" may be manipulated to provide immunization against disease. Genetic engineered agricultural crops, may be redesigned to yield only the most useful and valuable component desired—for example, orange juice sacs. Monsanto already has bio-engineered cotton to produce tinted cotton bolls on the plant. Bio-reactors take over where open field agriculture leaves off. Eventually, the same principles will be applied to clone human organs for transplant.

MEGA-MATERIALS ERA: DOMINANT BY 2200-2300

Mega-materials technologies, including the ability to deconstruct and reconstruct matter at atomic and sub-atomic levels to achieve desired properties, will radically transform the physical sciences ranging from architecture to an alchemist's wildest dreams. Advanced understanding of the bio-chemistry of life's instructional genetic codes, will be paralleled by analogous developments in physics and chemistry that involve quantum mechanics to construct "designer" materials. Harnessing nanotechnologies poses entirely new and novel echelons of mechanical interventions that previously were the stuff of science fiction. Fully understanding and adroitly manipulating the construction and deconstruction of atomic matter will require considerable time. Taking these technologies to a dominant position and role in the economy, may happen between 2200-2300.

Evolving discovery of atomic and sub-atomic structure

Step-by-step and part-by-part, a dithering array of particles and/or waveforms that make up atomic matter and structure continue to be discovered. An atom's nucleus occupies only some one-quadrillionth of its space. Peeling away the outer layers, like those of an onion, scientists have identified numerous sub-atomic components. The innermost depths and dimensions of sub-atomic matter remain unfathomed.

Evolution of imaging technologies

Manipulation of matter is inextricably bound up with the ability to "see" or image atomic and sub-atomic structure and processes. Advanced Photon Source equipment using X-ray probes can image the structural arrangement of molecules, proteins, and enzymes. Real-time imaging reveals chemical reactions as they occur. Once the precise nature of those sequences are understood, science will be able to control them at will.

Human ability to discern and to visualize ever small quanta of matter has steadily progressed over the centuries. These capabilities commenced with development of the simple microscopes. Advances in microscopic devices reveals smaller and smaller details of matter:

- —20X magnification with simple microscopes (1500s)
- —275X with the compound microscope (1590)
- —400X with first electron microscope (1931)
- —12,000X with advanced electron microscopes (1933)

—50,000X with high-voltage electron microscopes
—100,000 with electron micrscope (Zworykin, 1940)
—200,000X with X-ray microscopes
—250,000 with advanced electron microscopes
—1,000,000X with field ion microscopes
—24,000,000X with atomic force microscope (capable of imaging single atoms)
—>24,000,000 with scanning tunneling microscopes (1980)
—_____ with positron transmission microscope imaging (1998)
—50,000,000 with high-resolution microscope imaging (1999)

Miniaturization—doing more with less

Accomplishing desired effects with atom-thin layers, stretches out finite resource availability. Year-by-year, virtually any technology has been able—up to the point of limiting laws—to produce the desired effect(s) using less and less material. Miniaturization is a hallmark of technological progress.

Clocks, the forerunner of mechanical invention that eventually gave rise to the Industrial Revolution, initially occupied an entire temple. Pocket watches, ponderous and thick at the outset, steadily diminished in size. By 1700 the average thickness slimmed down to 1.5 inches, trimmed down to one-half that thickness (0.75 inch) by 1800, and shrank to 0.25 inches by 1850. Today, LED display timepieces mere wafers, some about the thickness of a sheet of paper.

The first programmable computer (ENIAC), filled an entire room. It weighed 30 tons, and measured ten 10 feet tall and 80 feet wide. Steam engines and internal combustion engines also were huge room-sized contraptions at the outset. Installation and placement, usually in a central location, involved a complex of gears, drive shafts, belts and pulleys that allowed remotely located equipment to tap into that power. Eventually motors became so small, powerful and cheap that miniaturized versions were embedded in each piece of equipment.

Computers are headed the same way as early massive engines. Birthday cards that tinkle the "happy birthday" tune when the card is opened, incorporate nearly the same computing power as the first computers. Palm-size computers and video games pack more computational capability than the best supercomputers of the mid-1970s. Motor vehicles with 30-100 onboard dedicated computers that constantly check and adjust oil pressure, fuel mixture, tire air pressure, seat adjustments, headlights, and navigation surpass the computational ability of Apollo-11, the lunar lander spacecraft!

Vacuum tubes shrank to peanut-size. Bulky tubes were replaced by transistors which, in turn, were miniaturized to sizes a fraction of a hair's width and densely packed into integrated circuits. Fiber optic cable, stretched thinner than a human hair, and eventually pulsed and encoded to carry hundreds of thousands of signals simultaneously across a broad spectrum of optical wavelengths, replaced huge tonnages of copper wires about one-fourth inch thick that initially carried but a single message. Unwieldy telephones of yesteryear have been replaced by featherweight versions with billions of times the capability as their earlier cousins.

NEW ATOMIC AGE: DOMINANT BY 2100-2500

Thermonuclear fusion, possessing the potential to satisfy unlimited energy needs, will usher in the New Atomic Age. Based on virtually limitless hydrogen, this key energy sector stands on the threshold of becoming another mainspring of advanced economies. Learning how to control the hydrogen-helium cycle that fuels Earth's sun, humans soon will master controlled energy extraction from their own "star furnaces." Thermonuclear technologies will dominate the economy when obstacles to controlling fusion are overcome.

Thermonuclear energy breakthroughs become urgent around the year 2050-2100, when petroleum resources dwindle and begin scraping the bottom of the barrel. Long before 2050, economies will become increasingly reliant upon coal. Then, around 2250-2500, when reserves of coal also dwindle and reach the limit of cost-effective recovery, the substitution of fossil fuels will be nearly complete, and the demand for alternative energy will become an imperative.

Hydrogen fusion energy sources on Earth, if developed, provide a prodigious energy supply that could last another 50 million years! The top ten feet of the ocean contain enough deuterium (heavy hydrogen) to supply projected energy needs on Earth for as much as 50 million years. Extraterrestrial sources will augment that supply, as necessary. The cost of converting hydrogen to a still more powerful form—tritium is not prohibitive. Deuterium-tritium reactions yield four times the energy output as deuterium-deuterium reactions. Beyond that, helium isotopes that provide a far more powerful energy resource than hydrogen isotopes, are likely to be exploited to meet any and all energy needs.

Petroleum

Recent production and consumption levels indicate 42-50 year supply availability of petroleum. American Petroleum Institute estimates of 1.4-2 trillion barrels which includes probable new discoveries and technologies are projected to provide supplies for 63-95 years.

Natural gas

Production levels indicate 54 year supply availability. This finite resource also is facing points of diminishing returns as supplies run out.

Coal

Provides some relief, at least temporarily. Production levels indicate 230 year supply availability, and up to 400-500 years with enhanced efficiencies. UNESCO estimates (c.1997) are more pessimistic, projecting that coal reserves will last only another 200 years.

Alternative energy sources

May provide some relief, but not much. Photovoltaic energy based on the sun's energy, pose prospects for supplies lasting for as more than five billion years—the time when sun's higher energy death throes and expansion snuffs out life throughout its solar system. Wind power, previously used to propel sailing craft and power windmills before the Industrial Revolution, is likely to increase somewhat. Although availability rose from 10 megawatts in 1980 to 7,630 megawatts in 1997, wind energy contribution remains relatively small in the overall energy supply. Fuel cells of an improved genre are expected to be commercially introduced by 2025. A variety of other minor sources also will contribute in a minor way to overall energy supplies.

Science, for over a half century has been able to manipulate less energetic fission reactions. This knowledge provides the brute force to unleash tremendous energy for destructive and constructive purposes. The threat of nuclear power, brought to fulmination over a half century ago, led to the abrupt termination of World War II. Since then, fission has been harnessed not only to create vast arsenals capable of nuclear annihilation, but scores of peaceful purposes as well. A growing number of nations are overwhelmingly dependent upon nuclear fission to produce and provide their electric energy needs. That list, and the percent of reliance, will continue to grow. The long and steady progress in developing nuclear energy has evolved over many centuries, with many more benchmarks yet to be established.

Current technologies can control atomic fission in a rudimentary way, but the extraordinarily high temperatures required for breaking the nucleus of hydrogen (100 million degrees Celsius for deuterium-tritium reactions), the difficulty of containing plasma, and raw material refinement pose obstacles yet to be overcome. Breakthroughs in plasma chemistry and physics, development of adequate containment fields (such as magnetohydrodynamics), triggering devices (such a laser implosion), refinement of raw material resources, and particle beam accelerators hold the keys to the successful development of commercially viable nuclear fusion. These obstacles are not insurmountable.

Beyond technological breakthroughs are the time and costs attending commercial introduction. Regulatory approval processes, construction, start-up, and actually bringing new energy sources on-line may require a minimum of an additional 20-50 years. This means that launch may begin around 2025, with substantial implementation accomplished by 2050. Keeping things in perspective, fifty years is not a long planning horizon.

NEW SPACE AGE: DOMINANT BY 2500-3000

Extra-terrestrial enterprise

Will become the main engine of economic activity sometime prior to the year 3000, perhaps as early as 2500. For centuries, increasingly sophisticated telescopes and observational probes have been searching and studying the far reaches of the cosmos. Increased magnification used to probe, visualize and image matter at sub-atomic levels steadily progressed to the point humans are on the verge of manipulating atomic and sub-atomic matter. At the opposite end of the spectrum of size, similar advances have been, and continue to be made, to the point that the conquest of the vast domains of the cosmos looms nearer than ever. Astrophysics and a score of other new disciplines and fields of inquiry slowly open up secrets of the universe. Spacecraft, manned and unmanned, already have begun the daunting task of exploring space.

Immensity of the cosmos is so great as to defy comprehension. The Milky Way in which Earth is situated includes 200-400 billion stars, many with numerous planets and large numbers of satellites. Beyond this is the immensity of the outermost reaches estimated to contain over 125 billion other galaxies! Despite vast dimensions, space visionaries already conduct businesses to arrange space travel for vacationers and thrill seekers, and burial of cremated remains in the cosmos. Others foresee space colonization, and resource recovery from the vast riches of this solar system, this galaxy, and beyond. This conquest will preoccupy Earth's attention into eternity.

Many of the New Space Age technologies providing the foundation and the means for pursuing extraterrestrial quests have been established. Now, these technologies and capabilities need to grow.

Forecasting capabilities

The future is not totally unknown. Exploring what the future holds requires diligent effort. There are trends and directions that can be discerned. There is some degree of probability about future developments inherent in the timelines of histories. Some clear notion of the economic "centers of gravity" that will dominate advanced nations over the next one thousand years can be anticipated.

THE CREATIVE UTILIZATION OF
HUMAN CAPITAL

CREATING TOMORROW'S DREAM TEAM

by

John Nance

(An Interview With Charles G. DeRidder)

Blending individual stars into a Dream Team of medal-winning performers is not just the goal of Olympic athletic coaches. Managers in business, industry, and government also strive to create teams of excellence. Now they can learn how from a unique training program of the US Forest Service designed to build teams that consistently perform at peak levels.

The secret?

To find out and focus on the natural strengths and mental preferences of each individual participant, and then to form these individuals into a cohesive team aimed at accomplishing a specific task.

Data gathered over the last three decades shows that teams organized with these principles consistently outperform randomly structured teams by from 25% to 65%.

These findings are derived from hundreds of seminars and team performances documented since 1969 in the Employee Development Program of the Pacific Northwest Region of the US Forest Service headquartered in Portland, Oregon. The program has been created, refined, and scrutinized by Charles DeRidder, Group Leader of Employee Development since 1968.

DeRidder, an engineer and forester who also has a doctorate in education, began with the belief "that existing or potential teams in most organizations could perform much better than they do."

Through years of trial, error, and continual personal observation and study—of the Forest Service and countless other small and large, private and public organizations worldwide—DeRidder put together a program based on time-proven and documented procedures. He fervently believes that by following certain basic steps, managers everywhere can increase team performances.

"This is no longer a fantasy," DeRidder contends, "you can make it real, make it happen."

His early experiences in various jobs, plus several years as a US Army Reserve officer, exposed him to a wide variety of training programs and opportunities. After joining the Forest Service, he started putting work teams together in the mid-1960s for Blake and Mouton's Managerial Grid Seminars. Since 1966, DeRidder's program has conducted more than 100 such seminars, ranging in size from

---

John Nance *is a photo-journalist, formerly with The Associate Press. Charles DeRidder is group leader of employee development at the Pacific Northwest Region of the US Forest Service, Portland, Oregon.*

four to ten teams per seminar, with each team consisting of six to eight persons. The goal was to train and enhance the performance of Forest Service managers from a variety of different professional disciplines. The teams were originally formed on a random basis with no boss-subordinate relationships.

In the late 1980s after reading *The Creative Brain* by Dr. Ned Herrmann, DeRidder saw great potential for utilizing Herrmann's concepts in the formulation of Managerial Grid teams. In the book, Dr. Herrmann lays out his findings that the human brain is specialized—not just physically, but mentally as well. This means that its specialized modes are organized into separate and distinct quadrants, each with its own perception, language, values, gifts, and ways of knowing and being. Herrmann wrote that each person is a unique composite of these differing modes according to a particular mix of mental preferences and avoidances.

In its simplest terms, Herrmann's four quadrants have key characteristics. A person operating in the Upper Left Quadrant (A) is a problem solver: mathematical, technical, analytical, and logical; a person in the Lower Left Quadrant (B) is more of a planner: controlled, conservative, administrative, and organizational; someone in the Lower Right Quadrant (C) is a talker: musical, spiritual, emotional, and intuitive; and finally an individual in the Upper Right Quadrant (D) is a conceptualizer: spatial, imaginative, holistic, and artistic. Herrmann's four quadrant model provides a basis for understanding the role of the brain in human behavior. It enables the array of many different thinking preferences to fit into a sensible whole.

Dr. Herrmann is quick to point out that there is a natural "tilt" in everyone, and he refers to this as "dominance." Dominance is a part and parcel of the human condition. He states that behavioral differences resulting from our mental preferences are, like handedness, perfectly normal forms of dominance. He notes that each person has dominant characteristics mentally as well as physically, and that our mental dominance ultimately affects behavior.

Not only do the left and right sides of our brain differ functionally, Herrmann says, but they are asymmetrical physiologically as well. The left hemisphere, for example, has a greater specific gravity, relatively more gray matter, and a wider occipital lobe. In contrast, the right hemisphere is heavier, has a larger internal skull size, and a wider frontal lobe.

Dr. Herrmann developed a battery of questions to determine the brain preferences of individuals who complete his survey instrument. This procedure, known as the Herrmann Brain Dominance Instrument (HBDI), provides DeRidder the framework necessary for understanding an individual's preferences. He took Herrmann's process and applied it to the formulation of grid teams.

DeRidder observes that by the late 1990s he had gathered sufficient data to be able to construct a variety of different teams—depending on the task they were to accomplish—with a high degree of assurance that the teams would perform considerably better than if he had put them together randomly or with limited or casual insight, as is commonly done. DeRidder said, "I can now structure Managerial Grid teams using HBDI concepts to produce better team scoring results than if I had put the teams together on a random basis as was done in the past. The data show that perhaps only one time in ten does a team that was formed "randomly" match the average of all the teams that I have put together using the HBDI. "

All the recent teams use the same pre-work, the same scoring devices, the same tasks, and the same seminar format for all sessions. The only variable is the way team members are selected and put together. Currently the team scores show a 25% to 65% improvement. "By any measure," DeRidder declares, "that is an excellent investment of your intellectual capital."

Findings show that the most successful teams are a composite whole brain (all quadrants equally represented within the team). This enables members of the team to:

- Understand and value their own uniqueness and differences;
- Understand and value others;
- Define their common purpose/mission;
- Commit to work together to achieve their purpose;
- Set specific performance goals;
- Measure progress;
- Agree upon team practices;
- Be accountable, and hold each other accountable.

Here's an example of how the process of forming a team works:

From a pool of nearly 60 participants, the task is to put together eight workshop teams—four teams of eight and four teams of seven participants. First, you want to have each thinking preference represented within the team by selecting the initial four team members based on each of them being dominant in a separate quadrant of the brain. Next the remaining team members must be chosen. Here it is crucial to consider the "task" that the team is going to address. Too often, this step is overlooked. But it is essential to consider the overall "mentality" of the task and the results you want to achieve. This helps determine the make-up of your work team.

For instance, if the task is to create a new work process or to improve productivity, you would next select a person who has a high preference for the "D" quadrant. Then your next two team members should have a good balance in all quadrants as they will act as interpreters for the entire team. For an eight-member team,

DeRidder's final selection would be an individual with a slight "tilt" to the left hemisphere for strength in practicality and analysis.

As we enter the next millennium, DeRidder says he is convinced that this process provides strong evidence that we can better utilize our intellectual capital. It will not only improve team performances on the job and thus increase productivity, but it will also enhance the quality of life for the individual. That is, if a person who has a low thinking preference for a particular task is placed on a job that requires high thinking preferences for that task, the individual's performance will be relatively low. Conversely, if the requirements of the job and an individual's thinking preferences align themselves, you have the potential of a world class performer who will most often give an outstanding performance.

DeRidder sees people working at their individual best who will find deep satisfaction in doing something they truly enjoy! When you see such people at work, he says, "it's obvious that they are happier, more relaxed, and more successful in the ways that count." Leaders and managers who are adept in using the brain preference process will be able to help their employees identify work opportunities that better match their personal brain dominance. Just possibly employees will cease to view their work as simply a "job." The potential now exists for the individual employee to pursue their life's work through their work life.

DeRidder says, "I'm convinced that the organization's intellectual capital base will be increased in value because of their enhanced wisdom, their judgement, and their inventiveness. I've seen it happen time and again."

Dream Teams of all kinds can become realities, DeRidder believes—gold medals are out there for the taking: "We are at a point in our development as an industrialized nation in which we have all the ingredients required to enter the next millennium and make our fantasies come true. The know-how is available, the key elements are in place. It's up to each one of us to decide whether we want to be a positive part of it. "

REFERENCES

Herrmann, Ned. *The Creative Brain*. The Ned Herrmann Group, 1993
Herrmann, Ned. *The Whole Brained Business Book*. McGraw-Hill, 1996

THE NEED FOR NEW PARADIGMS

by

John Diebold

No one knows the future but if we are to understand where we may be headed in the knowledge society we must look not only at today's capabilities and applications but at what I call the driving forces that are producing tomorrow's.

My colleagues and I have developed a model for the purpose of identifying specific opportunities produced by developments in information technology in the immediate future, through the year 2010 time span. While the model was created for analyses in connection with mergers, acquisitions, alliances and investments, I have chosen a few of the introductory frames to better describe the driving forces that underlie the development of these technologies.

FIGURE 1

MODEL FOR PURPOSE OF IDENTIFYING SPECIFIC OPPORTUITIES PRODUCED BY DEVELOPMENTS IN INFORMATION TECHNOLOGY

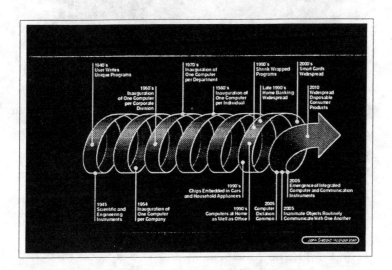

John Diebold *is the chairman of The Diebold Institute of Public Policy Studies, Inc., also the author of* Automation, Making the Future Work *and recently* The Innovators. *His office is in Bedford Hills, New York.*

To begin with, computers have gone from being viewed as scientific and engineering instruments 50 years ago through successive stages of what economists call capital goods; consumer durables and are already beginning to be used as disposable consumer goods—often mere by-products of services.

The principal points to focus on is that there are several driving forces:

- The technology itself has been the major driving force. It keeps forcing down the cost of the hardware and of transmission even as it increases system versatility and functionality.

FIGURE 2

DRIVING FORCES 1

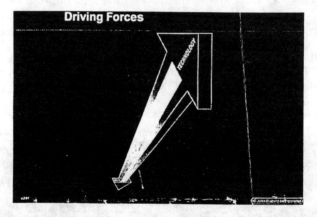

But—and this is critical—technology is not the only driving force. Others include:

- Increasing capacity and versatility of the technology
- Continuing price/performance improvement
- Resulting ease of human/machine interface
- Increasing importance of software
- Worldwide widening of computer literacy, expectation and use
- Societal factors
- Worldwide privatization and deregulation of the telecommunications industry—which are not necessarily a one way street

To understand the ways in which these forces interact, one must also study and understand what I call integrators that determine the interplay of the forces.

FIGURE 3

DRIVING FORCES 2

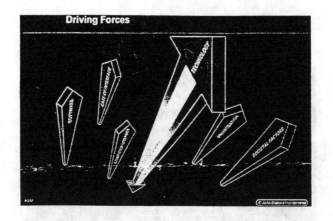

FIGURE 4

THE INTEGRATORS

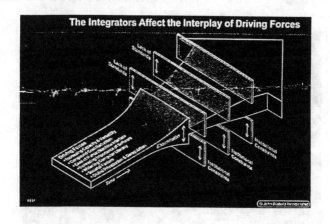

DISSEMINATION

- There are often preconditions to successive stages of development. One telephone is useless; additional ones create a valuable network. So it is with software, the Internet, etc.

FIGURE 5

INTERACTION OF DISSEMINATION, DRIVING FORCES, AND EVENTS

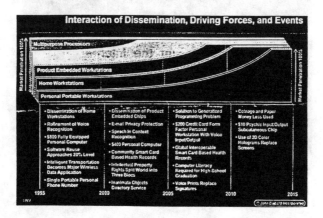

TIME

What doesn't work today may—and often does—tomorrow, due to:

FIGURE 6

TIME TRANSFORMS PREMATURE IDEAS INTO REALITY

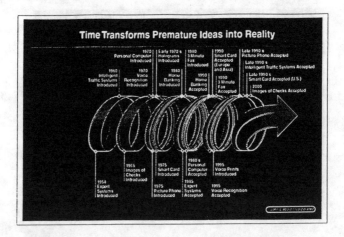

- Need to meet preconditions
- Where we are on the cost curve
- Economics of critical capabilities: wireless; voice; vision
- Innovations in software as well as hardware

STANDARDS

Interconnection

Overcoming impediments to effective economic demand (particularly in the public or non-market sector where the need is great but not yet the economic demand).

Miscalculation of integrators and their impact results regularly in business catastrophe for would-be exploiters of information technology who rely exclusively on technology and cost. Because cost and versatility make something possible is often a long way from having a product or service customers will buy!

From even the most cursory study of these forces and their interaction, it should be clear that phenomenal change is going to be taking place even in the immediate future to say nothing of the next half century. Easing the human/machine interface will itself radically change our daily lives.

THE NEED FOR NEW PARADIGMS

A century ago, when economists talked of the factors of production they meant land, labor and capital. In the late 19th century, we added management, or what business historian Alfred Chandler calls the *visible hand* on top of Adam Smith's *invisible hand* or market. In the 20th century services have overtaken physical goods as the product of production and of employment in more and more economies. Today intelligence or knowledge is emerging as sequel to "the great mass of data" put at our disposal by use of the information machines during the first 50 years' prelude to the knowledge society.

Each step of this long road has called for the working out of new paradigms in order to understand and manage our development.

So it is today, and rather than devoting the remainder of this paper to describing the many ways in which these new technologies are already changing our personal as well as work lives including:

- The widening role of the Internet
- Or the plethora of new enterprises (and accompanying millionaires) in Silicon Valley, Silicon Alley and other valleys, fens and glens in Europe, India and China
- Or the changing of the location and nature of where

we do our work
- Or our choice of career
- Or the difficulty of maintaining privacy
- Or the need for new skills.

I have chosen instead to describe some of the paradigm changes that seem to me necessary for us fully to benefit from the opportunities— and to avoid as many as possible of the negatives—as the knowledge society develops.

There are many examples of changes that have to be made from past paradigms, which have often served us well as our modern age has developed. The six examples I have chosen particularly interest me but many others deserve attention and study.

The ones I have chosen are:

1. New ways of organizing societal infrastructure
2. Widespread disintermediation as well as new forms of intermediation
3. The need to find an accommodation between very large global firms and smaller, entrepreneurial and human resource centered enterprises
4. Recognition that talent has become capital, and integration of this realization into our systems of management, organization, compensation and ownership
5. Changing our systems of education not only to play a central and continuous role throughout our lives but to utilize our newfound capabilities to improve learning and perception
6. Inventing ways in which private capital can be put at risk and entrepreneurial management encouraged to achieve new kinds of societal infrastructure.

NEW WAYS OF ORGANIZING SOCIETAL INFRASTRUCTURE

One of the most profound developments as the knowledge society comes into being will be the emergence of new kinds of societal infrastructure. The needs of the 21st century far exceed much of today's infrastructure and, fortunately, we have the means of being able to replace and reorganize much of our existing infrastructure, both what I call "hard," such as roads and bridges, as well as "soft," the systems for delivering social services, for example. What we do not have as yet are the public/private partnerships needed to attract the private capital and entrepreneurial drive to take the risks of innovation in creating the forms of infrastructure that are both possible and desirable.

For illustrative purposes, I have chosen six examples from several dozen more my colleagues and I have reviewed and which we feel to be important opportunities.

1. Information-intensive health care
2. Working at home
3. More easily arranged travel
4. Intelligent transportation systems (ITS)
5. Citizen access to government services and infor mation
6. Electronic shopping

As information technology increasingly changes our world, societal and institutional change becomes crucial to achievement of the full benefits which the technology makes possible. The examples I have chosen are important because they not only show how traditional businesses and infrastructure will be changed but how entirely new vistas are opened to a traditional business when it views societal change as representing business opportunities and is able to operate the services as well as build the systems. Thus information technology not only provides the means of responding to an increasing array of societal needs but it also plays a formative role in creating new business opportunities.

In each of these six examples of new forms of societal infrastructure, my exposition will follow the same pattern of organization. In order to succinctly show what I mean by a particular infrastructure I have used a schematic diagram of what I am suggesting. Then, in each case, I itemize some of the opportunities inherent in the depicted development, and then I list some of the obstacles to achieving the potential. I follow the same sequence in discussion of each of the six infrastructure— or infostructure—examples that I have chosen.

Examples of New Business Opportunities in Information Intensive Health Care

- Linking Health care sites in a metropolitan area creates systems integration and outsourcing opportunities as well as a transaction-based information services business.
- Claims processing systems shared by the major insurers will create major outsourcing opportunities.
- Both specialized and generalized information services companies will provide health care information to consumers and in some cases to health care providers as well.
- The market mechanism for patient records is not yet

- clear because the beneficiaries are different from those who incur the costs.

FIGURE 7

INFORMATION INTENSIVE HEALTH CARE

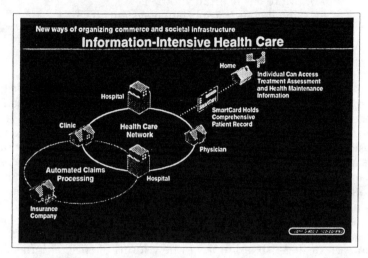

POTENTIAL

It is impossible to describe the full range of opportunities that this sea change in the way we organize health care will provide. In addition to the traditional opportunities for selling products to health care providers, there are major opportunities to build and run the systems that will link health care providers.

Much of the process of involving the individual in his own health care involves information and the processing of information. Much of this information can be provided by information service providers including those that will make their services available over the Internet. In some cases, individuals may wish to be able to use specialized software. Certainly the market for software to be used by health care providers and those offering information services to health care providers is a major growth area.

An important element in this process is the ability of health care providers, when authorized to do so by the patient, to access the patient's full medical history.

Obstacles to Realizing the Potential in Health Care

- Currently, computer processing of data is not reimbursable (in US).

- Market imperfection: Bearer of information technology cost and beneficiary are not the same party.
- Institutional issues: large number of players and rivalry among players.
- Privacy concerns.
- Licensure issues:
  a) Dividing line between providing information and practicing medicine.
  b) Provision of medical services from outside of jurisdictional boundaries.
- Infrastructure issues:
  a) How to fund infrastructure.
  b) Economic justification of infrastructure.

FIGURE 8

WORKING AT HOME

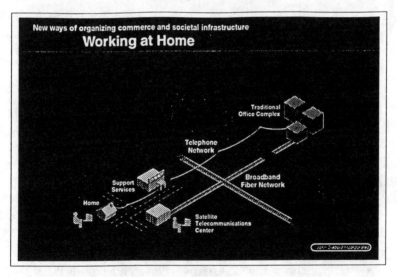

Examples of New Business Opportunities in Working at Home

- There is increasing demand for equipment, software, and new types of support services catering to homeworkers.
- Employers may require systems integration and other services to adapt their existing systems to the need of employees and contractors working at home.
- Geographic distribution of commercial real estate and support business may change substantially, particularly around major metropolitan areas.

- The trend makes it possible to attract qualified employees who prefer working at home and within their own time constraints.

POTENTIAL

This is a trend that is very important even though its full implications are still not fully understood. It is useful to consider a number of different categories of homeworkers. These include: self-employed and small businesses operated from home; employees of traditional employers who work from home on occasion; and employees who work from home in the evenings and weekends.

In the early 80s it was the conventional wisdom that the technology to facilitate homeworking was already available. Now we realize that there are certain specialized requirements for homeworkers and as these are incorporated into various products, we would anticipate an acceleration in what already is a fast growth rate.

Obstacles to Realizing the Potential in Working at Home

- Inadequacies of software and services to make working at home transparent to others.
- Inadequate bandwidth at home (ISDN)
- Lack of acceptance of concept and conflict with those who do not participate.
- Confusion over the status of homeworkers:
  a) Tax disincentives for self-employed
  b) Liability issues.

Examples of New Business Opportunities in Travel

- Generalized information service providers have the opportunity of obtaining a large share of the travel booking business now provided by specialized travel service providers.
- Specialized travel services will survive only if they cater to niche markets or are able to provide something extra.
- Prepaid smart cards will play an increasing role in local transportation.

Examples of New Travel Related Services

Beyond pretrip travel information and handling bookings:

- Proactive intervention when a planned itinerary is at risk.
- Message service since the travel computer knows

the itinerary of the traveler.
- Currency and other financial services.
- New networks linking travel agencies with travel suppliers, the multitude of tour operators, a variety of other information of interest to travelers as well as major corporate clients.
- Managing travel expenses. Services provided to an organization might include:
  a) Analysis of total organization's use of travel expenses
  b) Oversight of individual employee misuse of travel
  c) Poor decisions linked to frequent traveler programs
  d) Less than lowest cost travel
  e) Other patterns needing review.

FIGURE 9

MORE EASILY ARRANGED TRAVEL

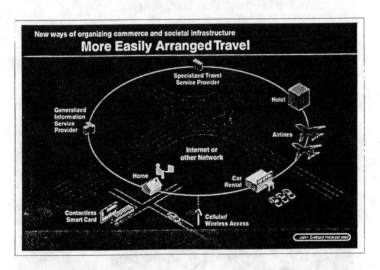

POTENTIAL

Travel is presented here because it is representative of many other changes that are taking place. Existing players (i.e. the specialized travel agents) will be displaced, find safe niches, or partner with more generalized information services providers and prosper.

Travel will be one of the human activities driving the trend towards the replacement of cash by electronic money.

Travel will resemble health care (i.e. all of the travel providers will be electronically interconnected and also interconnect to the travelers).

In some sense this is already the case today with the major airline reservation systems. But these systems are not generally usable directly by the traveling public, and this will be one of the major changes taking place. Also the current airline reservation systems are relatively closed systems. Clearly a major hotel chain will be part of each airline reservation system and the travel agent will be able to book a room there at the same time as the flight is booked. But today, every bread and breakfast in the world is not included in the Sabre system. Tomorrow essentially every B&B will be on the Internet.

Obstacles to Realizing the Potential in Travel

- Payment mechanism particularly for those without access to traditional credit cards
- Tracking system for travelers who wish to be able to be located
- Initial cost to develop content and establish commercial arrangements
- Standard content and transaction formats.

FIGURE 10

INTELLIGENT TRANSPORTATION SYSTEMS (ITS)

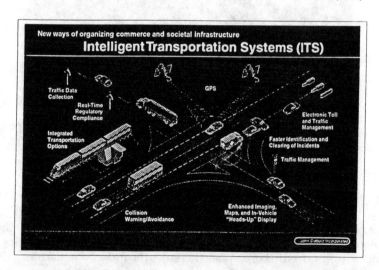

Examples of New Business Opportunities in Intelligent Transportation Systems (ITS)

- Prepaid transit cards may provide an entree into the more generalized electronic purse marketplace. A similar opportunity exists in education and other specialized communities of interest.

- In-vehicle safety and navigation equipment is becoming a large market.
- Travel information and roadside assistance/Mayday services can be provided by specialized and generalized information services providers.
- Heavy demand for wireless communications will result from the above. Wholesale purchase and resale opportunities exist for service providers and municipalities.
- Traveler information, as opposed to pure routing and congestion avoidance, provides an opportunity to participate in the revenues resulting from motel, restaurant, and entertainment reservations.

POTENTIAL

In the United States, approximately 30 distinct ITS and services areas have been identified. A European or Asian analysis would be very similar since transportation requirements are the same everywhere although priorities and important details vary.

The 30 United States defined ITS user services translate into roughly 80 distinct market packages which may be purchased and deployed. A market package is a set of functionality which can be deployed and which has benefits in excess of the costs.

For each of these 80-odd market packages there is a value added chain with respect to production, distribution and after-sale support. There are a variety of roles a company can play particularly with respect to services. Thus, the array of possible participations is both extensive and complex.

Obstacles to Realizing the Potential in ITS

- Privacy Issues
  a) Individuals
  b) Fleet operators and individual drivers
- Bearer of cost and beneficiaries may be different parties
  a) How to fund the infostructure
  b) How to pay for safety investments
  c) How to pay for energy and clean air investments
- Equity concerns: How to maintain public support without destroying the business opportunity
  a) Ratio of free to paid services
  b) Ratio of benefits to paying versus subsidized drivers
- Conflict between increased mobility and energy and

clean air priorities
- Uncertainty about the size and timing of the market opportunity

FIGURE 11

CITIZEN ACCESS TO GOVERNMENT SERVICES AND INFORMATION

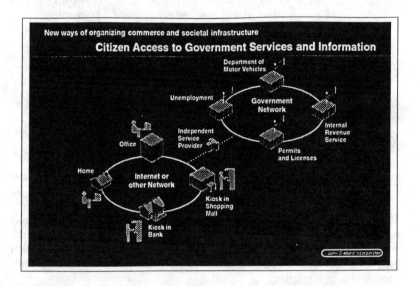

Examples of New Business Opportunities in Citizen Access to Government Services and Information

- New mechanisms for transfer payments will become very important as will personal identification products and services.
- Banks will be major players.
- Kiosks at unmanned locations will also play a role.
- The need for cross-agency computing creates opportunities for systems integrators and outsources even as it creates bureaucratic problems.
- Privatization of certain categories of citizen/government transactions is likely.

- Privatization of the marketing of governmental data is likely when privacy is not an issue.

Obstacles to Realizing the Potential in Citizen Access

- Privacy concerns
- Institutional issues related to:
  a) Role of the private sector
  b) Cooperation among governmental organizations
  c) Federal/state/local
  d) Multiple jurisdictions
- Availability of cash alternative systems
- Availability of personal identification systems including ways to obtain and share the database information particular to each individual
- Funding considerations:
  a) Based on savings to government
  b) Based on value added to citizens and businesses served

FIGURE 12

ELECTRONIC SHOPPING

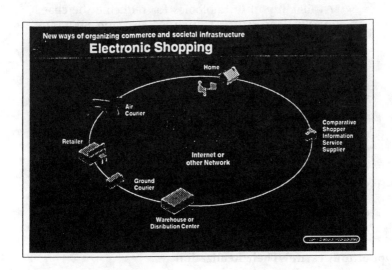

Examples of New Business Opportunities in Electronic Shopping

- Information services to allow for comparison shopping

- Software agents to reduce the cost of on-line access
- Increase in geographic reach of successful retailers
- Opportunity for every manufacturer to now be in mail order business
- Increasing demand for package delivery services
- Added volume for credit card companies
- New payment mechanisms required for those unable to obtain conventional credit cards.

POTENTIAL

Many categories of companies will be impacted by electronic shopping, all but a few impacted in a positive way. In addition to the direct impacts of increasing the volume of products delivered via package deliverers rather than being self-delivered by the purchaser, and the ability of those with a product to sell to greatly expand the geographical reach of their distributions channels, this trend is creating a demand for a wide range of information technology products and services to facilitate electronic shopping and these represent major opportunities.

Obstacles to Realizing the Potential in Electronic Shopping

- Secure payment mechanisms
- Availability of technologies for reducing the chaos involved in current Internet use for shopping purposes:
  Directories
  User friendly interfaces
  Software agents
- Current lack of software and content to enhance the appeal of products which cannot be touched, smelled, seen full size or in true 3-D
- Communication costs if extensive searching and comparison shopping is involved
- Product delivery and return costs
- Reactive actions by retailers and employees negatively impacted
- Times: electronic shopping will inevitably become more important

Opportunities and Their Realization

Historically societal change of this scope results in extraordinary business opportunities.

Most enterprises continue to operate as they did previously, finding it more and more difficult. Typically, a few entrepreneurs recognize the opportunities, regroup their resources and play the new game by the new rules and thrive.

While consumer and societal infrastructure markets are only embryonic at this point, they represent enormous growth possibilities. I have tried to show in these six examples the pervasiveness of the ways in which information technology will change our ordinary world and work, as the cost/performance ratio and the technology makes it increasingly comfortable for ordinary untrained people to interface with complex systems.

I would like to make clear that in choosing these six examples, I am not implying that they are necessarily the most attractive societal or business opportunities in information intensive infrastructure. There are many others. For example: monitoring and responding to environment change; security; municipal and public administration. Indeed, in building the model from which these examples were extracted, my colleagues and I explored and incorporated 36 categories of opportunity of which the six I have cited here as well as other forms of infrastructure represent but one! But I hope that they serve to illustrate one of the major categories of change possible as the knowledge society develops.

In many ways the public sector examples I have cited are among the most difficult to effect since they require major institutional changes. I cite them because even though they are very large, they are so easily overlooked precisely because they represent and require major societal changes; we are not yet used to thinking of building and operating societal services as business opportunities. The following chart may be helpful in suggesting how these phenomena are likely to develop.

A crucial problem is that most of the important activities of our governments—particularly those that primarily manage or operate rather than set policy—have been operated as monopolies largely removed from markets. Recent moves to benefit society through privatization have been denationalization. Creating competitive markets in areas that have been considered public service monopolies rather than business opportunities is a big change for the managements of these enterprises. Joseph Schumpeter's "winds of creative destruction" can be very chilly for some years in such situations!

In trying to change this, one runs up against a morass of laws, regulations and obsolete organizational structures. It soon becomes clear that the risk-averse nature of the bureaucracies which operate these institutions takes a long time to change. Gaining acceptance for new ways to meet the changing needs of society is a monumental task. But on all grounds, from benefits to society to creating new business and high paying job opportunities, it is worth doing. But there will be many difficulties in bringing these opportunities to fruition. Yet the changes I have outlined will certainly take place.

I would like to make clear that I am not opposed to government *per se*. What I am opposed to is government operation of such things as the delivery of public services. At the same time, I believe government should focus its effort far more diligently in proactively setting public policy rather than reacting to changes for which it has not prepared.

FIGURE 13

MARKETS ARE AT EARLY STAGES OF DEVELOPMENT

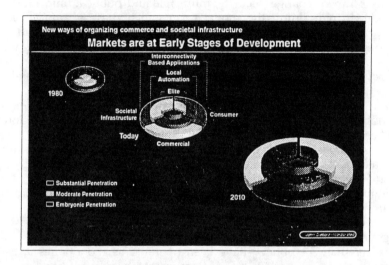

Widespread Disintermediation as well as New Forms of Intermediation

Information technology is gradually changing the way products and services of all types are sold and distributed. Today, new ways are operating alongside the old. Ultimately, the new way will in many cases supplant the old.

1. The need for multiple levels of distribution is quickly vanishing in industries where human interaction currently plays a routine role in determining the flow of prod-

uct or service. For example, when any of the hundreds of Procter and Gamble products is run over the scanner in any Wal-Mart store, the information flows directly to Proctor and Gamble computers which manage their inventory. Replenishments are sent to each store with no intervention by the store or chain managers. Wal-Mart never even takes ownership of any Proctor and Gamble products that are sold in its stores. A true example of disintermediation. The potential for increased economic efficiency is very large, eliminating what are often several layers of intermediaries.

2. The Internet provides a departure point for consumer-oriented electronic commerce. It has triggered a wide-spread behavioral change made possible by the dissemination of communicating PC/workstations. The addressing system established by the Internet protocol coupled with the changed behavior of users and the wholesale disintermediation it is creating are perhaps the most important and enduring aspects of the Internet itself. The concept that all home and business computers are interconnected has very quickly been turned into a reality.

3. Voice recognition, after a number of false starts, is extending the realm of the information industry to new categories of untrained, non-technical users. Both voice recognition and the less technology-challenging (but also less user friendly) voice response and menu selection techniques allow automation to replace or supplement humans at many levels in the distribution chain.

4. The Need for Intermediaries

Case Example: Management of External Information

The manager of external information represents the interests of the enterprise with respect to external information. The manager of external information:

- Arranges the procurement of external information
- Installs and operates the information systems necessary to support cost effective search, fusion, distribution, and use of external information.
- Establishes appropriate systems and policies to assure external information is being used cost effectively.
- Avoids duplicate purchases of information and oversees

the disposal of unnecessary and obsolete external information.

- The user can perform the management of external information function internally or outsource these functions to a third party. Third parties will be able to provide a more sophisticated and comprehensive service than internal managers and that is why we believe that this represents a business opportunity.

5. The ability to integrate products from multiple suppliers provides many advantages to the end users, but it also creates the need to initially integrate and continually maintain the integrated products. Most users do not possess these skills in-house.

6. There is an increasing need for global information and communications systems as global customers seek to develop integrated systems.

7. There is a large installed based of information technology products that can now be upgraded by software and product components rather than being replaced.

8. There is a need to manage information as well as equipment and personnel.

9. A market is developing in software components (sometimes called "objects") which can be purchased individually and combined to build a useful software system. In turn, software produced by users may have a value if it is developed as free standing interconnectable components (objects).

10. How These Intermediaries Interact

- Systems Integrators
  a) Specify and then purchase equipment and software from vendors.
  b) Design and contract for or internally produce additional specialized equipment and software where required.
  c) Take responsibility for ensuring that all of the purchased elements and specially designed units are able to work together to perform the desired functions.

- Data center/client server/and regional network outsourcers.

- Assume responsibility for the day-to-day operations and maintenance of: (1) Information systems and (2)

Communications systems.
- Generally perform some functions normally assigned to a systems integrator when these are not of a scale justifying a separate procurement.

FIGURE 14

COMPLEXITIES CREATE NEED FOR INTERMEDIARIES

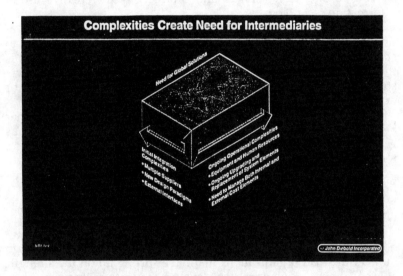

Global network intermediaries

- Procure communications services from communications providers (1) in every geography where customer has major operations, (2) from communications providers representing the mix of alternative communications capabilities required and (3) perform the systems integration of these capabilities to the extent that this is required.
- Provide a single point of contact for the customer to avoid the customer having to deal with multiple providers and provide a single invoice covering all communica-

tions services required with the ability to present these charges in any way that is meaningful and useful to the customer.

The Need to Find an Accommodation Between Very Large Global Firms and Smaller, Entrepreneurial and Human Resource Centered Enterprises

The ability to find and manage an accommodation between very large capital-intensive, global firms and smaller human resource-intensive, entrepreneurial enterprises will increasingly determine success not only in the information industry but in other industries striving to offer new services and products by the creative use of information technology.

One aspect of the rise of high-tech entrepreneurism, where innovative employees flee the security offered by big companies in order to start risky ventures of their own based on their creative strength, is the new interplay between the entrepreneurs and the giants via strategic alliances' investment positions; marketing agreements; or outright acquisitions.

For example, the number of alliances of one form or another between small bio-tech firms and large pharmaceutical firms has grown from 58 in 1993 to 327 by 1996, and has continued since. Something like a third of mergers and acquisitions are already information and communication based transactions, mostly large firms acquiring smaller ones.

A dual market structure is developing in a number of industries -- with a small number of increasingly large global players with world-class manufacturing and distribution facilities, international know-how and massive financial capabilities on the one hand and a large number of small organizations that have sprung up on the strength of innovations. As new markets develop, they interest the giants just as the entrepreneurs are finding it increasingly difficult to develop as global firms. One of the major problem/opportunity challenges in today's business world is finding managerial accommodation between the giants and the infopreneurs.

Interestingly enough, the "webs" that are starting to be a prevalent way of coping with this situation and with the strategic alliance problems, may well provide help for the problem of monetization of intellectual capital. They provide large return to the creative people, but still keep them at arm's length from highly defined (and confined) personnel practices of the giants.

- Sector I companies seek Sector II innovations, creativity, growth rates and high return on net assets (RONA).
- Sector II companies need world-class global resources as they become mid-sized.

196

FIGURE 15

CONTINUATION OF "DUAL" OR
BIFURCATED INDUSTRY STRUCTUE

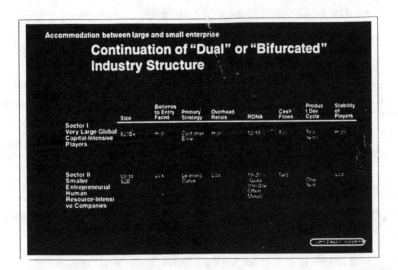

- As Sector II companies succeed, they become of increasing interest to Sector I companies, but have their attractions all too easily destroyed through mergers or badly handled alliances.
- Sector II companies planning to grow and transition into a Sector I player, rather than seeking an accommodation with the Sector I players, will find it difficult to succeed.

Need for Accommodation Between Sector I and Sector II Companies

- There are many differences between the culture and natural approaches of Sector I and Sector II companies.
- Increasingly, Sector I companies foster Sector II operations in order to participate in high growth areas.

- As Sector II companies grow they face two problems:

- Their mature products/services take on the characteristics of Sector I businesses.
- Increasing size leads to the problems previously identified as the Dual Industry Dilemma. What seems so easy when the company was a $200 million operation and which still seemed doable as a $1 billion operation becomes more difficult above $3 billion.
- A high growth rate soon means adding enormous revenues each year.
- The CEO can no longer be the product manager.
- Sector I companies that used to see the Sector II companies as an ally with respect to other Sector I companies now want to reclaim what they see as a logical market for themselves.

Alliances between Sector I and Sector II players logically will be beneficial to both parties but historically have been problematical to one or both parties.

- IBM/Microsoft
- IBM/Rolm
- AT&T/SUN Microsystems

PROSPECTS FOR THE CONTINUATION OF A DUAL INDUSTRY STRUCTURE

- Factors arguing for an end to the dual structure:

  a) Threats to the ability of Sector I players to maintain and protect customer bases
  b) Demise of proprietary systems
  c) Breakdown of national/regional protected telecom markets
- Threats to the natural advantages of Sector II companies if the production of software becomes less human-resource intensive

- Factors arguing for a continuance of the dual market:
  a) Increasing importance of established distribution networks
  b) Global alliances if they in fact are successful
  c) Significant innovations continuing to come from Sector II companies
- What are the prospects for important Sector II players?

Mergers, Acquisitions and Alliances are Often Required

- New roles may require competencies which may be

difficult to rapidly grow internally.
- Where being global is important, mergers and acquisitions can be helpful, but alliances may be better and far less expensive if they can be managed.
- Where increasing product volumes are required to amortize increasingly large development and tooling costs, it may be necessary to acquire or merge production with competitors to gain the required scale.
- The rapid growth of new product/service categories means that mergers, acquisitions, and alliances are necessary to participate in these opportunities. Internal development will generally be too slow and lack some elements of success attainable through mergers, acquisitions and alliances.

A premium will be placed on understanding how alliances need to be structured and managed particularly when they involve both small and large companies of quite different characteristics including the case that one is technologically based and the other not.

Recognition that Talent has Become Capital, and Integration of This Realization into Our Systems of Management, Organization, Compensation and Ownership

Little more than lip service is normally given to the obligatory annual report line that the enterprise could not have achieved all it has without the talents and hard work of the employees. But already in today's world we can see with increasing frequency case after case where one or two individuals determine the success or failure of an enterprise or a product or service. The problem is that most of our personnel management systems and our accounts were set up on the idea of democratic treatment of a homogenized workforce.

We have no accounting method (other than "good will") for recognizing what is almost always our greatest asset. Yet even this early in the emergence of a knowledge society it is obvious that, as Louis B. Mayer, proprietor of MGM, used to say, "In this business the assets go home every night."

It is human creativity that is in reality the driving force of the knowledge society. Yet we largely lack the tools for properly managing creativity. For example, one of the key factors affecting the outcome of the current telephone company/cable/computer/motion picture wave of mergers will be the extent to which the key players understand that managing creative talent is quite different from running a telephone monopoly and that unless they find a way of working with managements who are good and seasoned at the new game, they may find they have bought very little in their acquisitions.

It is entirely possible that the models to which businesses look in developing managers are less appropriate than they once were. More appropriate organizational, personnel and motivational models might be:

The Large Movie Companies that provide financing, marketing and distribution for constantly changing coalitions of creative people who produce the films.

Sports clubs, record companies and TV that understand talent is capital and treat it accordingly—individual by individual.

Research organizations that have a success record such as the Cavendish Laboratory under Lord Rutherford, (considered by many as the "most productive decade and laboratory in the history of science"), the Xerox Palo Alto Labs (from which the innovations of much of Silicon Valley have come, but from which Xerox has failed to benefit until very recently)!

Organizations such as the Advanced Research Projects Agency (ARPA) of the US Department of Defense with a shining success record of backing innovators and fledgling technology based ventures.

We can probably learn more about building effective organizations for today and the 21st century by looking at non-conventional models than we can by applying the personnel and management principles of the past.

Increasingly, the businesses that succeed are those that recognize talent as capital and find innovative ways of bringing the best talent together whether through alliances, subcontracting, free lance arrangements, or any number of electronic linkages that allow people to shift time as well as geography in their own manner and preference in working. The ability to manage these enterprises is surely closer to the *Ballets Russes* than to G.M.

For two decades, 1909 to 1929, Diaghilev ferreted out the best of then largely unknown talent in art, music and dance. He produced a series of ballets that continue to be performed today and have changed the face of the dance world. He arranged his own funding and changed the market and consumer behavior in his field. He did this by forging and managing artistic and commercial alliances that brought together composers such as Igor Stravinsky, Claude Debussy, Maurice Ravel and Sergey Prokofiev; painters such as Pablo Picasso, Andre Derain and Henri Matisse; and choreographers such as Michel Fokine, Vaslav Nijinsky, Leonide Massine and George Balanchine. The brilliance of Diaghilev's creative management in finding, stimulating and coordinating these largely unknown talents was fully as creative as the talents he marshalled.

Is this not more nearly the model required to manage as well as create many of tomorrow's enterprises? It is already the case in both California's Silicon Valley and Manhattan's Silicon Alley. To carry it off requires new thinking.

Today, talent is capital, but it is not yet perceived to be by most large corporations, their accountants nor by our business schools. Yet is not the ability to find outstanding and perhaps unknown talent, attracting it, and managing the often extremely tense interpersonal relationships the key to so many of today's -- and especially tomorrow's -- enterprises? This is the case, whether they are multimedia or software producers on the one hand or the rest of industry -- from finance and transportation to publishing and travel -- that are all being changed by these new technologies.

Changing Our Systems of Education to Play a Central and Continuous Role Throughout Our Lives and to Utilize Our Newfound Capabilities to Improve Learning and Perception

The one thing that can surely be said about education at the moment, in every country with which I am familiar, is that most people are dissatisfied with their present systems!

There is a generally accepted mantra that education must be lifelong rather than considered complete when schooling ends early in life. But it is hard to find any place where that is happening in an acceptable way.

It is equally difficult to find more than a few attempts to tap the obvious revolutionary potential of using information technology, to say nothing of biotechnology, to change the entire educational experience. We must certainly ask ourselves why it is that when computer technology is literally redefining so many industries, as yet no sign of it is being utilized—in primary schools at least—other than providing some PC's, in arguably the most important function (education) in both our societies? Why is this so?

My own belief is that by removing education from the private sector (since it was considered too important to leave to the--heaven forbid!--private arena), we effectively removed it from Schumpeter's turbulent winds of creativity and innovation.

There can be no question that new paradigms in education are essential even to pretend that we are becoming a knowledge society.

None of the potentials I have been discussing will be achieved if we do not radically change our approach to education. After all, the "knowledge" in "knowledge society" must come from education, just as wisdom comes from the experience of applying knowledge. But we are a long, long way from knowing what these new paradigms must be.

Since even relatively trivial parts of our consumer society account for far better use of technology than in education, which is infinitely more important, the principal thought I have is that we should focus some effort on overcoming the impediments to producing effective economic demand for successful innovations in education.

In the United States I am happy to see that some encouraging innovations bubbling up from citizen groups and, in a few hundred cases—so far in elementary schools—have overcome the reactionary forces of the educational establishment. Evolving in quite different ways in different parts of our country, vouchers, charter schools, for-profit schools, new kinds of nonprofit enterprises all attempt to move decision making to the parents as to what school a pupil attends—providing more than one alternative.

Educational innovation does not necessarily have to be brought about through the market but choice is clearly the element that is currently missing in education. Might this not create conditions which would encourage new approaches? It is worth a great deal of experiment.

Inventing Ways in Which Private Capital Can Be Put at Risk and Entrepreneurial Management Encouraged to Achieve New Kinds of Societal Infrastructure

If the kind of advanced infrastructure needed in the 21st century is to be achieved, it is going to be necessary to take risks, both technological and, usually more important, financial risks—risks dependent upon what the market will accept in usage and transaction charges for advanced services. For many reasons, some of them quite good, the public sector in all countries is risk averse. (One might well say that much of German society is risk averse.)

The surest way to attract the best entrepreneurial and management talent to tasks, such as creating new forms of infrastructure, is to establish a risk/reward ratio that will attract private capital. You can be sure that the best management talent will accompany such investments.

Despite all the attention to privatization (which, worldwide, has primarily meant denationalization), finding private capital for societal objectives is a considerably harder task on which to focus public policy. The Diebold Institute partnering with Deutsche Bank have launched a fellowship program to underwrite research on cases, both success and failure, in such endeavors. Fellowships will support promising mid-career Europeans to spend a year in America and the reverse for American fellows—each to complete a comprehensive case study.

It is some measure of the novelty of the idea that we have so far had a relatively small number of applicants on each side of the Atlantic for what are lucrative fellowships—though I must admit we are searching for truly outstanding individuals. But public/private partnerships are essential if we are to succeed in meeting our infrastructure needs in the 21st century. Both sides of such partnerships are essential for success, the innovation of the private sector and the societal acceptance of the public sector.

CONCLUSION

The six examples of the need for new paradigms are cited only to illustrate my point that bringing about progress toward a knowledge society will require considerable change in our theories, the forms and processes of both our public and private institutions, and in many more aspects of the way we live and work.

There are many other examples that I might have chosen and that certainly should be studied, such as:

- **Changes in Our Political Processes**

  It is hard to believe that our political process will not be changed as we develop as a knowledge society.

- **New Tools and Measures**

  Similarly, new paradigms will emerge in the concepts and theories by which we organize and govern our societies.

- **Society's Legal and Ethical Underpinnings**

  Similarly, public policy, legal and ethical problems abound in moving from our present state to a knowledge society.

Vetoes and Priorities

- In our admirable desire to ensure democracy in all our decision processes, we have created literally hundreds of places where we can thwart initiative with a veto.

Longer Time Horizons

- All too often our time horizon is short-term, while the key to solving many of our problems is long-term vision and commitment.

In this present moment of history in which:

- Physicists report that a first form of teleporting is possible—a precursor of the "beam me up" of *Star Trek*?
- the US National Football League is launching a chain of electronic entertainment centers
- new categories of scientifically designed

food—nubricenticals are already a staple in some peoples diets
- new categories of businesses are making profits delivering new services dealing with everything from the way we entertain ourselves to how we make war upon one another
- the convergence of bio-technology and info-technology is beginning to be seriously discussed.

Can it be doubted that new paradigms are called for?

To understand, much less create, the needed paradigms is an effort that should rank high on public as well as private agendas. It needs the serious attention of a great many wise and skilled people from philosophers and scientists to public administrators, politicians, political scientists, educators, financiers and poets. Much, if not everything, will change—including some things as basic as the way we view ourselves and our role in the world and, quite possibly, the physical form of mankind. It is, indeed, hard to imagine any human institution that will not change.

I should like to place before you a recommendation. There could be created an institute for the continual assessment of the human and societal consequences of technological change so that some possible alternative futures could be imagined and discussed well ahead of finding ourselves with pressing problems rather than a smooth transition, thus providing a proactive rather than a reactive view of our needs and resources as we move into the age of the knowledge society.

Such an institute might be national at first, but surely should be planned along international/universal lines:

- It could act as a research clearinghouse. Interchange agreements could be maintained with the hundreds of institutes and research centers producing work that throws light upon the kinds of problems and opportunities that await us. It need not do fresh research itself but would make a major contribution as a clearinghouse or database.
- It must not be exclusive, but inclusive.
- It must not discourage the duplication of effort, but foster as many attempts to find answers to as may questions, or variations of the same questions, as possible.
- It must be a center not of authority, but of light.
- It must provide focus, but leave no area of investigation deliberately dark.
- It must be structured and managed in such a way that it does not inhibit innovation, but encourages political and societal innovation adequate to insure we encourage and

make use of innovations so society benefits from them and not discourage them.

Above all, such an institute should try to assess not only the immediate and obvious consequences of research, technology and related developments but the impact of accelerated change itself.

THREE PARALLEL REVOLUTIONS: MINDING THE ECONOMIC IMPERATIVES OF KNOWLEDGE

by

William E. Halal

Adapted from the author's book, *The New Management: Bringing Democracy and Markets Inside Organizations*

Berrett-Koehler, (1998)

In late 1997, Bernard Ebbers, CEO of WorldCom, a small, obscure firm in Mississippi, announced that he was buying MCI for $42 billion of his company's stock. It was the largest takeover in history. How could this unknown man, a former gym teacher, emerge from nowhere with no capital to seize control of the second largest telecommunications company in America and gain immediate dominance over the global communications market?

Ebbers forged this empire with little more than a keen understanding of how a jumble of diverse companies could be integrated to deliver a complete stream of communication services around the world—a task that eluded AT&T, MCI and foreign telecom giants. Because he grasped the underlying insight needed to create this system, all else followed.

Countless other examples show that knowledge is today the most powerful force on Earth, primarily responsible for the collapse of communism, the restructuring of economies, and the unification of the world. After decades of glib talk about the Information Age, companies are becoming "learning organizations," developing their "intellectual assets," and hiring "Chief Knowledge Officers" because we now see that this is the source of all productivity, innovation, and competitive advantage. It is suddenly, blindingly clear that knowledge is a boundless source of infinite power that promises to flood the world with creative progress. Bill Gates told a group of CEOs that information technology will "fulfill their wildest dreams."[1]

THE CONTRADICTIONS BETWEEN CAPITAL AND KNOWLEDGE

The problem, however, is that this vast divide between a limited past and a boundless future has left business adrift in confusion—the flavor of the month management fad syndrome—because we lack what economists call a workable "theory of the firm" for a knowl-

---

William E. Halal *is Professor of Management, George Washington University, Washington, DC.*

edge-based economy. The "Old Management" of the Industrial Age is dying because it was based on capital-driven economics, and we now know that enterprise is no longer powered primarily by capital. Former Shell executive Arie De Geus says "The critical resource is now people and the knowledge they possess."[2] This means that most corporate practices of today no longer make sense for the world we are entering.

Corporations comprise economic systems that are as large as entire nations, yet it is a great irony that our most admired companies remain committed to roughly the same type of central-controlled hierarchy that failed in the Soviet economy. We have seen a few marginal changes, but the bulk of useful knowledge lies unused among employees at the bottom of the firm and scattered outside its walls among customers, suppliers, and other groups—while most decisions are made by executives at the top.

This yawning gap between promise and reality merely hints at the enormity of the upheaval that lies ahead. The entire social order is being uprooted by the move from a capital-centered past to a knowledge-centered future—even while we remain confused about what to do, where this is going, and what it all means. Without a theory of the firm based on the logic of knowledge, today's struggle for survival will remain an endless exercise in bewildering change and management fads.

A NEW FOUNDATION BUILT ON AMERICAN IDEALS

I want to suggest that a well-established foundation for a "New Management" of the Knowledge Age is readily available if we would simply look in the right place. America's heritage of democracy and free enterprise could serve us exceedingly well in this new frontier. Unfortunately, we tend to relegate these ideals to political elections and competition between firms. But if managers could extend the liberating power of democracy and markets *inside* business corporations, government agencies, and other social institutions that govern the daily flow of ordinary life, their widespread use would have a profound impact.

This is not some hopelessly utopian quest because I intend to illustrate that trends are moving rapidly in this direction.

To survive a world of constant change, massive diversity, and intense competition, leading corporations are dissolving into fluid collections of self-managed units that use local knowledge to carve out successful market niches. As I will show later, the necessity of this bottom-up approach should bring the power of enterprise to fruition as organizations melt into a churning sea of "internal markets" offering all of the creative dynamism of external markets—call it "the flowering of enterprise."

The move to democracy is equally apparent in the way creative managers work closely with tough competitors, empowered employees, and discriminating clients. After a long history of conflict, collaborative working relations have become one of the most powerful forces in business because of the hard realization that the mutual sharing of knowledge with other parties is beneficial. Some companies such as GM Saturn are uniting their stakeholders into complete "corporate communities" think of it as "the extension of democracy."

If managers could take a fresh look at these rich but misunderstood trends from the perspective of our traditions, the emerging pattern could guide our way ahead with confidence. *As this book will demonstrate, the power of democracy and enterprise promises to transform institutions for a new era.*

Why should we be surprised? This is the philosophy that gave birth to the United States and that has brought down dictatorship after dictatorship. Free markets and democratic governance are the twin pillars supporting modern civilization. They are proven methods that we have found most useful because they involve us all in making decisions that govern our society.

THE COMING PARALLEL REVOLUTIONS

This article describes leading-edge concepts and practices derived from my study of the successful experiences of progressive companies. It's a strategic plan, or a guidebook, designed to help us figure out where all this is going.

Follow me through the examples and ideas here and you'll learn about three parallel revolutions that make up this transition to knowledge-based organizations. Figure 1 sketches out the flow of revolutionary advances along three major paths: 1) the Information Revolution that is driving this transition 2) the resulting transformation of business, government, and other institutions 3) the creative new forms of leadership emerging to handle all this change.

Note that these trends follow a rising exponential curve that is characteristic of all change today—the typical "J curve" depicted in the figure. Whether it is the number of computers in use, strategic alliances, or new ventures, the trendline is curving sharply upward.

THE TECHNOLOGY REVOLUTION

Just the Beginning of Unstoppable Change

Thus far we have seen only the first rumblings of the IT explosion that is yet to come. The simple stuff is over and the most innovative, wrenching innovations lie ahead. I conduct a forecast of technological advances every two years, and the latest study detailed the arrival of 85 revolutionary breakthroughs.[3] This wave of technological change is poised to crash over society during the next few decades as the rising power of IT feeds back to improve IT itself. Technology is basically knowledge, and the widespread use of IT is now driving our understanding of technical knowledge at ever faster rates. Here's a rough timetable of three major breakthroughs:

1) 2003 +/- 2 yrs Interactive multimedia should be used by people everywhere to work, shop, study, and conduct all

FIGURE 1

PARALLEL REVOLUTIONS IN TECHNOLOGY, ORGANIZATION, AND LEADERSHIP

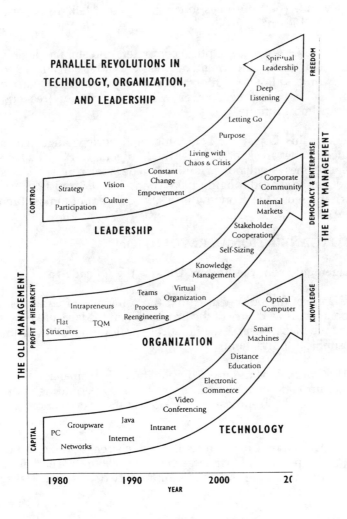

other activities electronically over life-sized wall monitors. Electronic commerce is expected to reach $12 billion by the year 2000 alone.

2) 2009 +/- 3 years Smart machines, robots, and software should be able to interact with people, learn and reprogram themselves, and translate languages. Bill Gates said: The future lies in computers that talk, listen, see, and learn."

3) 2014 +/- 4 years Optical computers and storage devices (as in the Superman movies) should be available to process limitless information in any form. Andy Grove said: "Computer power will be practically free and almost infinite."

In short, this is just the beginning of historic change that seem destined to alter all aspects of life. The IT of today—PCs, the Internet, cellular phones—will look primitive in a decade or so. The U.S. stock market has advanced roughly 1000% between 1985 and 1999 because Americans sense the economy is entering an era of almost limitless progress.

THE ORGANIZATIONAL REVOLUTION

Management From the Bottom-Up and the Outside-In

The heart of my argument is that the principles of enterprise and democracy form a theory of the firm based on the laws of knowledge —*The New Management*. Two heretical applications follow from this philosophical foundation:

1) **Internal Markets** Complexity is best managed not through planning and control—but by permitting widespread entrepreneurial freedom at the bottom of organizations.

2) **Corporate Community** Economic strength flows out of not power and firmness—but the collaborative exchange of knowledge among the community of corporate stakeholders.

TOP-DOWN CONTROL DESTROYS THE BULK OF CORPORATE WEALTH

During the 1990s decade of Capitalism Triumphant we constantly heard about the evils of central planning and authoritarian control, but anybody in business will tell you that the prevailing corporate system remains a centrally-managed hierarchy adorned with a few gentle touches and good intentions. Despite fervent claims about empowerment, networking, teamwork, and other hot management concepts, this has also been a decade of harsh downsizing, top-down change, and extravagant executive pay.

For instance, IBM's Louis Gerstner may have pulled Big Blue back from the brink, but only by reinforcing fierce discipline and hierarchical control. IBM managers described their new boss this way: "His blunt style sent tremors through the organization." In 1997, the value of IBM's individual divisions totalled $115 billion while the parent company was valued at $65 billion; the missing $50 billion was consumed by corporate bureaucracy. IBM's managers claim the software division alone wastes $200 million each year getting headquarters approval of its 10,000 software projects. [4]

Meanwhile the shock therapy approach to restructuring has become a way of life in America—even though it is now notorious for meager economic gains, overburdened staffs, badly served clients, and alienated employees. In 1999, for example, GE's John Welch was planning to close plants, sell divisions, cut wages, and lay off thousands.

This top-down approach may work in the short-term but—like painting over rotted timbers—it masks the underlying weakness and invites catastrophe, as we've seen in the decline of AT&T, Sears, GM, and many other former corporate giants. Top-down management is not going to withstand the massive changes looming ahead as relentless hypercompetition drives open a frontier of new products, markets, and industries that nobody really understands. Andrew Grove, CEO of Intel, put it best, "The Internet is like a tidal wave, and we are in kayaks." [5]

Downsizing, for instance, seems to make sense from a capital-centered view, but the knowledge held by employees comprises 70% of all corporate assets! [6] To put it more sharply, the economic value of employee knowledge exceeds by far all of the financial assets, capital investment, patents, and other resources of most firms. Firing people is akin to throwing the bulk of corporate wealth out the window.

Downsizing can be best understood as a palliative, ritualistic practice, akin to bloodletting in primitive medicine. It has become an icon of our time that reveals a far more serious organizational illness. Corporations shed workers repeatedly because they suffer from a chronic inability to create growth in a confusing new economic frontier. Instead, they downsize as a bad habit, providing temporary relief by reducing labor costs while actually draining energy as they lose skilled workers, creative ideas, loyalty, and other vital assets.

INTERNAL MARKETS RELEASE KNOWLEDGE FROM THE BOTTOM

The solution is a fundamentally different approach that harnesses the creative talents lying dormant in average people. While Fortune 500 dinosaurs downsized by three million employees during the 1990s, smaller firms and new ventures upsized by creating 21 million new jobs. This salient fact shows that the key to vitalizing organizations is to bring the liberating power of small enterprise inside of big business.

In short, we need to shift the locus of power from top to bottom, to think of management in terms of enterprise rather than hierarchy. I know this sounds revolutionary, but this *is* a revolution as dramatic as the Industrial Revolution. We tend to hear the *Information* half of the Information Revolution but ignore the *Revolution* half. Just as the idea that Communism might yield to markets seemed preposterous a few years ago, similar change is needed in big corporations—"Corporate Perestroika." Robert Shapiro, CEO of Monsanto, put it this way, "We have to figure out how to organize employees without intrusive systems of control. People give more if they control themselves."[7]

Hundreds of companies have used clever forms of internal enterprise to solve problems directly, creatively, and quickly. Pay-for-performance plans are being expanded to form small, self-managed units that are held accountable for results but free to choose their workers, leaders, strategies, work methods, and generally "run their own business." Line and support units are being converted into profit-centers that buy and sell from each other and from outside the company, converting former monopolies into competitive business units. MCI, Xerox, Johnson & Johnson, Hewlett-Packard, Motorola, Siemens, Lufthansa, and other companies have developed fully decentralized bottom-up structures that form complete "internal market economies."[8] ABB's 4,500 independent profit centers stand out as a model.

It only takes a little imagination to extend these trends to the point where the logic of free markets governs rather than the logic of hierarchy. The concept of internal markets has profound implications because it shifts the source of knowledge, initiative, and control from top to bottom, thereby providing the same benefits of external markets: better decisions through price information, customer focus, accountability foe economic results, and as much entrepreneurial freedom as possible.

Yes, markets are messy, but they are also bursting with creative energy—roughly like the Internet, our best model of self-organizing market systems. Nobody could possibly control the Internet's complex activities, yet by allowing millions of people to pursue their own interests, somehow the system grows and thrives beyond anything we could imagine.

In the final analysis, only a new form of management based on enterprise can meet the explosive challenges lying dead ahead. The hope that "participation," "team spirit," "inspiring leadership," and other vague ideas can create dynamic action among tens of thousands of people in the typical organization is little more than pious wishing. Anyone who has ever managed knows that it is almost impossible to get more than 20 people to agree on anything. Mayor Steve Goldsmith of Indianapolis told me that he struggled for years trying various management methods, but nothing worked like turning his departments into self-supporting units competing with outside contractors.

Here are three simple but bold actions that highlight sure-fire ways to jumpstart your organization:

1) **Link Resources to Performance**

   Rather than use budgets and other crude controls that are unrelated to results, link resource allocations to economic and social value created by units.

2) **Allow Units Total Freedom**

   Allow all units almost total operating and strategic freedom, including the right to buy and sell from partners both inside or outside the firm.

3) **Replace Downsizing with Self-Sizing**

   Let units handle their own staffing rather than impose layoffs. That is, "self-sizing" instead of downsizing.

Why would tough minded executives yield control over these crucial matters? Because they can thereby lead an organization where everyone shares the responsibility for success.

THE PROFIT-MOTIVE DESTROYS THE POWER OF SOCIAL PURPOSE

This does not mean that CEOs give up power or that corporations are balkanized into warring camps. The role of executives shifts to designing these self-managed systems and providing leadership to unify diverse interests into a strategic whole, a "corporate community." Saturn, The Body Shop, IKEA, and scores of enlightened companies develop trusting relations with clients, share power with workers, and cooperate with suppliers, while also making more profit for investors.   It's important to stress that these companies are not simply "doing good." They create value by pooling knowledge among stakeholders to solve management problems. In other words, corporate community is economically effective.

Beyond the sheer benefits, however, lies a vast and more powerful world of meaning and purpose. Corporate community is also essential to help us find our way through a turbulent world engulfed in an avalanche of expanding information. It is a great paradox that having more data often leaves us more *confused* out of the sheer limitlessness of it all. We are beginning to understand that information is meaningless without being guided by relationships, values, and vision—all those subtle but very real qualities lying beyond knowledge.

Unfortunately, these things run counter to the ideology of capitalism. The traditional idea that corporations owe their allegiance to shareholders and profit places managers in an unrealistic position where they are opposed to the interests of employees, customers, and others whose support is essential. Employee pay and training, for instance, are then viewed as simply costs to be avoided. But the more complex reality is that both employee welfare and profitability are perfectly compatible. Companies that form employee partnerships enjoy huge returns on their investment in labor.[15]

Consider how the health care industry provoked the public's wrath by cutting patient services to improve profits. Congress passed laws banning such practices, and 2000 physicians called for change because HMOs are "destroying the soul of medicine."[9]

How did a great profession dedicated to serving humanity get into such a mess? In pursuing today's notion of good business, HMOs lost sight of their social purpose. It's obvious that we must control costs and investors must be rewarded, however, any business must serve society to survive.

This business-society conflict has everyone confused, wasting energy rather than working together toward common goals. Robert Haas, CEO of Levi Strauss, explained the problem, "People look through the wrong end of the telescope, as if profits drive business. Employee morale, turnover, consumer satisfaction ... that's what drives financial results."[16]

CORPORATE COMMUNITY DRAWS KNOWLEDGE FROM OUTSIDE GROUPS

If American executives can look beyond the bottom-line, they would find vast opportunities for profitable business flowing directly out of joining with the interests of their stakeholders. In the health care industry, for instance, progressive HMOs are involving all parties in collaborative problem-solving to improve health care while reducing costs. Typically, physicians, nurses and other staff are organized into self-managed practices that are accountable for

214

performance but given wide freedom and support. Education programs assist patients in managing their own health better and in *preventing* illness by adopting healthier lifestyles. And to keep the system honest, states provide access to medical performance data to let market forces work. Doctors are now often stunned to see patients show up with a clutch of medical research reports in their hands.

Here we see the power of knowledge-based enterprise. Progressive HMOs are redefining medicine into a more effective system of collaborative problem-solving among administrators, medical staff, patients, their employers, and government—corporate community in action. This approach allowed Oxford Health Plans to double in size each year to serve 1 million members, [10] and other companies in every industry could make a similar transformation.

But doesn't this compromise the need to make money? A knowledge economy is changing the old assumption that profit and social benefits are opposed. Unlike the fixed limits of capital, knowledge *increases* when shared, which is why cooperation has now become efficient. For instance, today's wave of strategic alliances is fueled by the pooling of technology, market access, and other forms of knowledge to increase value for all partners. Ray Smith, CEO of Bell Atlantic, calls it the principle of loaves and fishes: "unlike raw materials, knowledge can't be used up. The more you dispense, the more you generate."[11]

If cooperation can multiply the value of alliances with business partners, why shouldn't it be effective for *social* alliances with employees, customers, and other groups? Results reported from my "Corporations in Transition" (CIT) survey of 426 managers show that more than 80% accept the need to collaborate with all stakeholders.[12]

Although I like this idea because it resolves the age-old clash between business and society, I do not argue this case on moral grounds. Corporate community is not social responsibility or business ethics—it's one of the few remaining ways to sustain competitive advantage.

Imagine how the following creative but tough actions would electrify your organization with fresh knowledge from the outside parties you depend on to succeed:

Democratize Corporate Governance

Invite responsible, well-informed representatives of employees, clients, and business partners to sit on the board of directors and other bodies.

Evaluate Financial and Social Performance

Develop measures of performance that reflect the contribu

tions and benefits of all stakeholders, as well as traditional financial performance.

Collaborate Among Stakeholders

Use this democratic form of governance and performance measures to engage all stakeholders in joint problem-solving to improve the overall system.

These changes are not a luxury but a necessity for any business that hopes to meet the test of social purpose. The disorders of our time represent a vast frontier crying out for a new type of enterprise that creates value by integrating different interests to serve all needs better.

THE LEADERSHIP REVOLUTION

Relinquishing the Illusion of Control

If I am right, organizations are heading toward some sort of "economic reversal"—a passage from hierarchy to markets and from conflict to community. We seem to be roughly half way through this passage, and the principles of a New Management are quietly gathering momentum. Exploding complexity is forcing decentralized controls, while the benefits of collaboration are attracting diverse parties into pockets of shared understanding.

The way ahead seems clear. To manage organizations in a new era when ordinary people offer the most valuable resource available, leaders will have to push authority down to the bottom and out to all affected parties—a New Management based on shared leadership from the bottom-up and the outside-in. My CIT study shows that managers generally understand this shift is coming, and they expect it to arrive between the years 2000-2005.

It is certainly needed. Today's creative destruction of free markets is uprooting the old social order, posing mounting potential for a serious economic backlash. The income gap between the top and bottom classes in the US has returned to levels prior to the Great Crash of '29, exceeding all other industrialized nations, and indices of social well-being have reached new lows.[13] And much more turmoil lies ahead because industrialization is likely to increase 10-fold. China alone will triple the use of scarce resources, global competition, social diversity, and pollution.

MOVING THROUGH THE PASSAGE: LEADERS AS GARDENERS

I suspect the only way this conflict can be resolved is by *moving through the passage*—by harnessing the potential of a New Management based on the laws of knowledge. The key is to see that capitalism is dying but enterprise and democracy are just beginning to flower. To realize these possibilities, however, leaders have to relinquish the illusion of control to adopt a more humble but realistic role of *nurturing* rather than commanding their organizations.

The new science of complexity and chaos theory shows that organizations today must become shifting clusters of self-controlled autonomous units, a living superorganism of countless small cells that constantly adapt to a turbulent world. The Old Management was good for mechanistic business, but the New Management asks executives to give up their old role as captains of commerce to become "economic gardeners" of organic systems.

I experienced this coming role shift when attending a fundraiser at my son's high school recently. These used to be loud, hard-sell events that auction prizes to the highest bidder, leaving people dazed but feeling sort of loyal for attending what was basically an unpleasant bash. This time we were invited to enjoy a quiet dinner with a few other parents and teachers at small tables. The result was a meaningful dialogue over the raising of our children and the role of the school. Rather than leave with my normal headache, this fundraiser produced a deeper appreciation for this institution I entrust my son to. And the school benefited not only from our heightened support but from the more generous checks we willingly wrote after an enjoyable encounter that left us all connected.

It seems to me that this is what leaders have to cultivate today. The glitzy marketing, brutal treatment of throw-away workers, and all the other relics of a more exuberant but thoughtless economic youth must yield to a maturity that is quieter but more powerful. Leaders must find a way to serve unmet social needs, develop information systems to sharpen our understanding, help employees organize themselves into self-managed units, and form collaborative relationships to resolve the old conflicts between workers and managers, sellers and buyers, and all the other divisions we can no longer afford.

Leaders can't force people to do any of these complex tasks any more than gardeners can force nature to produce what they want. Gardeners are attentive to the subtle signs of need in their garden. They must provide the right amounts of water, light, and nutrients and then lovingly allow plants to grow as they should. In other words, they must let go. Listen to how Bob Kuperman, CEO of Chiat/Day, described this new role, "Basically our organization is now a living thing with a life all its own. Management can support it and guide it, but not control it. If you let it design itself, it takes off and people use their best possible abilities. We've got to make this succeed because the old way doesn't work anymore".[14]

HERITAGE, HERESY, AND THE LAWS OF KNOWLEDGE

One particularly crucial, symbolic action would signify these three revolutions, help us grasp them, and live up to the challenge. Drawing on our heritage as a nation born through revolution, Americans should summon up our traditional courage to proclaim a modern heresy—our economic system should no longer be thought of as "Capitalism."

Capitalism is an outmoded type of market system dedicated to the pursuit of capital, profit, and the other material factors that worked in the industrial past. The main thing holding us back at this point is sheer ideology. If we want to draw on the energy of the future, we should define our economic system in terms of the laws of knowledge that define the future. Economic success is no longer powered by capital but by free enterprise and democratic community. I suggest a more accurate, fitting name would be "Democratic Enterprise."

Corporate executives are the primary candidates for creating this system because business is the most powerful institution in society. A system of democratic enterprise would allow us to more easily navigate this economic passage, and managers could then shed their old role as the bad guys to assume their rightful place as the heroes who make a knowledge society work.

REFERENCES

1. *New York Times*, May 10, 1997.

2. Geoffrey Colvin. "The Changing Art of Becoming Unbeatable," *Fortune*, (November 24, 1997).

3. William E. Halal. "Emerging Technologies," *The Futurist*, (November-December, 1997).

4. "Defending Big Blue," *Newsweek*, (September 30, 1996). Betsy Morris, "Big Blue" *Fortune*, (April 14, 1997).

5. Grove is quoted in "A Conversation with the Lords of Wintel," *Fortune* (July 8, 1996).

6. Thomas Stewart. "Trying to Grasp the Intangible," *Fortune*, (October 2, 1996).

7. Robert Shapiro. "Growth Through Global Sustainability," *Harvard Business Review*, (January-February, 1997).

8. William E. Halal et. al. *Internal Markets*, (NY: Wiley, 1993).

9. David Hilzenrath. "Doctors Lash Out Against Profit Motive," *Washington Post*, (December 3, 1997).

10. "Oxford's Education," *Business Week*, (April 8, 1996).

11. William E. Halal (ed.). *The Infinite Resource*. San Francisco: Jossey-Bass, 1998.

12. William E. Halal. *The New Management*. San Fransisco: Berrett-Koehler, 1996.

13. *1996 Index of Social Health*. Tarrytown, NY: Fordham Grad Center, 1996.

14. William E. Halal. *The Infinite Resource*. Jossey-Bass, 1998.

FUTURIST OBSERVATIONS ON A NEW MILLENNIUM

AFTER THE PARTY IS OVER: FUTURES STUDIES AND THE MILLENNIUM

by

Graham H. May

With the possible exception of 1984, the year 2000 has been the major focus of attention for Futurists for much of the last fifty years.

From, at least, the 1960s on, a series of studies were published, either with 2000 in the title or as their focus (Appendix 1 lists some of these studies). The year 2000 has therefore been an important concept and image in our work for almost as long as Futures has existed. As we approach that landmark our vision has, as good Futurists, quite logically moved ahead to 2010 (Northcott 1991) 2025 (Coates, Mahaffie and Hines 1997) and beyond, but as a result of the 2000 centered work we have probably the largest collection of Futures material focused on one particular time ever accumulated.

How will we and the public in general react to this? Assuming that much of this writing is still available we may expect considerable media interest in the period leading up to the millennium and immediately after. Under the headline "Futurologists aren't what they used to be," *The Sunday Times* reviewed the predictions made in 1850 of John Such in a spoof copy of *The Times* for January 6,1950. (White 1999) Noting that he did get some things right much of the article focused on predictions that did not come true. Particularly where they are accompanied by visual images, which always look quaint, this "Ho-Ho" factor is a great way to sell newspapers, magazines and television programs. We should be prepared then, for some difficult moments, as we are asked, "What went wrong?" and "How can you justify your activities as a Futurist when you cannot get it right?" As Futurists we are well aware of the difficulties of prediction and frequently contend that it is not our business to forecast the future, but to help people prepare for its uncertainty by considering possible, probable and preferable alternatives. To us, that is eminently reasonable, based as it is on our professional understanding of our relationship with our futures, but from the outside it can appear as an excuse for incompetence coupled with a plea for continued support in the face of apparently damning evidence.

As Futurists, the focus of our attention now is forward into the 21st Century and the new millennium. Reasonably so, for that is why we exist; but it is important that, because we know that the future is uncertain, we take a learning approach to our endeavors (May 1997).

Graham May *is principal lecturer of futures research, School of the Environment at Leeds Metropolitan University, United Kingdom.*

One of the resources we have to assist our learning is the accumulation of experience in the form of the studies of the year 2000. Nobody knows better than Futurists that the past is a less than perfect guide to the future, but if we are to advance our trade and develop our skills it is likely that there are lessons to be learned from these efforts. A valuable research project into the history of Futures and the lessons to be learned awaits. It is only to be hoped that the resources are still available and have not been pulped to avoid their authors blushes, or because it is assumed their value is past.

In the run up to the millennium interest in both the past and the future has been growing. At the end of each year the media run several reviews of the year and usually rather fewer forward looks.

At the end of each decade a similar but longer period is reviewed and speculated about. Few alive today will remember the turn of a century but it is no surprise that the end of a century and the beginning of a new millennium is creating many times the interest.

In the United Kingdom, although there is considerable skepticism and many have expressed the opinion that the money could be better spent, large sums have been expended on the Millennium Dome. Just down the River Thames from the Greenwich Meridian, where of course, the Brits think the Millennium will really begin, the Dome is marketed as the major national focus of celebration. Throughout the country though there are other activities and developments, many of which have been funded by grants from the Millennium Commission, itself, supported by money from the National Lottery. The City of Leeds in the north of England is, for example, redeveloping a part of the *centre* as the Millennium Square. Other countries, though perhaps not on the same scale are planning similar events.

It may be the result of subsequent generations failing to benefit from the experience of their predecessors, the cyclical pattern of human affairs, or the imminence of the millennium, but there is evidence of a resurgence of interest in thinking about the future. After the enthusiasm of the 1960s Futures, and long-term thinking in general, in the UK and US particularly, went through a relatively lean patch during the late 70s and 80s. The laissez-faire doctrines of Thatcherism and Reaganomics were not sympathetic to the kind of forecasting and planning common in the 60s and early 70s. Much of the strategic planning effort of that period was seen as part of the problem rather than part of the solution. Forecasts, which had often been based on little more than simple extrapolation, were shown to be inaccurate and the market was deemed more responsive and flexible than any plan. If it was neither possible to forecast accurately nor make plans that remained relevant for more than a short time it was considered a waste of effort to indulge in them.

The 90s have seen a return of interest, but one with clear distinctions from earlier efforts in the 60s. In business the confidence in strategic planning has been replaced by an awareness that the future does not neatly develop along predictable tram lines, but remains unpredictable with numerous possible futures contingent on factors, only some of which any individual, firm or government can influence. In response to this understanding the preferred technique of the 90s, in business at least, appears to be scenario planning, which contends that the best way to prepare for the uncertainty we face is through the consideration of alternative futures. Numerous expensive workshops and seminars, a growing army of consultants and a number of books indicate the trend (Schwartz 1991, Van Der Heijden 1996, Galt et al 1997, Ringland 1998). For Futurists this is a welcome development as we have been canvassing the value of scenarios since the 60s. Just one word of caution, however. As Futurists we need to use this opportunity to learn and evaluate the technique or as with other "business fads" (Clarke & Clegg 1998) scenario planning may soon be cast on the scrap heap for promising more than it can reasonably expect to deliver.

Governments have also shown renewed interest in thinking about the future, particularly in the area of research, science and technology. Following the example of the Japanese; Finland, Germany, Hungary, the United Kingdom, Canada, Nigeria, South Africa, India, Korea, Singapore, Australia (Office of Science and Technology 1998) and New Zealand have all embarked on Foresight Exercises in the last ten years. The United Kingdom Technology Foresight Program was launched in 1993 as part of a government drive to encourage British industry to become more competitive through the development of science and technology. A steering group and a series of industry focused panels produced reports in 1995, which have set priorities for government funded research. Subsequently the program has been renamed Foresight and a second aim, enhancing the quality of life, added. The second round of Foresight to be launched in mid 1999, though still focussed on science and technology has a wider scope with main themes considering the ageing population, crime prevention and manufacturing 2020. Sectoral panels concerned with particular parts of the economy are retained but there are in addition two underpinning themes: education, skills and training and sustainable development. Details of the program can be found at http://www.foresight.gov.uk.

Welcome though these exercises are, they are often limited. It has been suggested in respect of the UK program that, to misquote Neil Armstrong, "It is a small step for man, but a giant leap for the British Government." The focus of these studies is usually science and technology and its exploitation in the interests of the competitiveness of the national economy. Some do go further but their location, usually within Science and Technology Ministries, raises some concerns that other themes, such as the quality of life and environmental issues, may be considered secondary to the major focus. Technology seems likely to be a major driver in the 21st century but how it is used, by whom and for whose benefit are crucial questions that require equally serious consideration.

The appeal of the future at the turn of the millennium is also apparent in the new publications that are being launched and the column inches in established journals. Several Futurists were invited to contribute to the special edition of the American Behavioral Scientist devoted to Futures Studies in Higher Education edited by Jim Dator (Dator 1998). His introductory paper contends that most readers of the journal will be unfamiliar with the discipline of Futures even though it has existed for over 30 years while Wendell Bell sees a preferable future in which Futures Studies becomes a major force within academia and particularly in the social and policy sciences. In the United Kingdom the new journal *Foresight* was launched in early 1999 as, a year earlier, were a number of popular magazines, including *Frontiers and Tomorrow's* World, the latter based on a long running BBC Television series. Elsewhere The Sunday Times (1999) published a five part series in association with the Millennium Experience, the company running the Greenwich Dome, entitled Chronicle of the Future. Modelled on the Chronicle publications it presented "tomorrow's news today" from 2000 - 2049. It is likely to be the first of many, as in the run up to the millennium there is an increasing flood of future-focused writing and television.

The quality of this material is varied. The best is thoughtful and thought-provoking, encouraging the kind of debate about futures that Futurists have been arguing for; the worst is appalling and has the potential to do serious damage to Futures activity. Much of these outpourings will be published for the sole purpose of maximizing sales of the newspaper or journal concerned and will frequently tend to the sensational. The "Gee-Whiz", "Shock-Horror" and "Giggle" factors may be expected to feature prominently and it is likely that accusations may be made that Futurists are a sinister group attempting to control the future. "There's a group of people who want to control your future. They think they know what's best for you.... These people describe themselves as 'futurologists' or 'futurists'. Be very afraid" (Gardiner 1999). Neither the author, nor Ian Pearson of BT, the two futurists/futurologists featured in the article, think of themselves in those terms, rather, as founder members of the UK Futures Group they are concerned to increase the debate about the potential futures, but that is not as eye-catching.

Getting ahead of the game also has commercial advantage. This is not new. Several writers, Naisbitt, Toffler and Popcorn, to name a few, have established track records in this field. Their work has not been without criticism. Richard Slaughter (1993), for example, in "Looking for the Real Megatrends," contends that such work is usually culture bound and lacking in critical analysis. Again we may expect a great deal of what Slaughter dismisses as "pop-futures" around 2000. Much of this is short-term, marketing based material, useful in its own right perhaps, but frequently confused with longer term Futures work.

As we approach 2000, the Y2K, or millennium bug, is bound to attract more attention. Opinion before the event is divided. Some contend that although there may be some disruption and that it might be wise not to be in the air at the stroke of midnight on December 31st 1999, it will not be a significant problem. At the other extreme there are those who regard it as a major threat to civilization. We may expect their warnings to become louder as 1999 proceeds, particularly if some of the precursors of the main event, such as the 9th of the 9th 99, brings the computer crashes that some predict.

In all this activity there is a resurgence of Futures as an academic discipline and a serious endeavor on an international front, but it remains small and underfunded. It is , for example, interesting to note in respect of the Foresight programs, that there appears to be an underlying assumption that anyone can do it. In one sense this is true, one of the characteristics of being human, it is sometimes suggested, is our ability for mental time-travel, both to the past through memory and to the future through our imagination. As Futurists, however, we know that although thinking about the future is easy, to be valuable, such thought needs to be grounded in careful analysis and an understanding of the limitations of the processes involved. All that seems to be required by many of the Foresight programs is an expert, scientist, historian or technologist and they can turn on the foresight. That there might also be a skill to thinking constructively about the future, that needs to be developed, just as there is a skill in the study of history, seems to be ignored. We need to be aware then that many experts will be trotted out to predict the future, and though few will have experience in Futures, they will be regarded as such.

Before we get carried away in our enthusiasm for the millennium we, in the West, also need to remember what it is we are commemorating. To western nations with a Christian tradition the millennium has a significance which is not mirrored in other cultures. The millennium is firmly based in the Christian calendar, which to us is normal, and quite literally, every-day. It is this calendar which informs us that the year 2000 occurs at this particular time but the calendar is a human construct, which closely approximates, though not exactly, astronomical phenomena. It is also related to our decimal counting system, if we used base 12 instead of 10, as some have suggested we should, it would not be the year 2000. Even in its own terms, 2000 years since the birth of Christ, it is generally thought to be incorrect, as that event is usually held to have occurred some years BC according to the very calendar we use.

Other cultures have different calendars and celebrate their new years at different times. There is, therefore, no exact match between the year 2000 in the western calendar and the year of other cultures, but roughly 2000 equates to:

5119 in the Mayan great cycle
1420 in the Muslim calendar
1992 in the Ethiopian calendar
1716 in the Coptic calendar
2544 in the Buddhist calendar
5780 in the Jewish calendar and
6236 in the first Egyptian calendar
2749 in the Babylonian calendar
 208 in the French Revolutionary calendar

Not only do we have the prospect of alternative futures, but alternative presents as well.

Globalization has carried the western tradition far beyond its Middle Eastern origins and European heartland. First, Europeans colonized and attempted to convert much of the rest of humanity to their ways, including their religion. More recently the New (to them) World, as the Europeans termed it, in the form of the United States has carried the western perspective further and deeper into other cultures, but paradoxically, although economically western ways may be dominant for the time being, culturally there is a resurgence of the local. (Naisbitt 1994) This is, in a number of examples, a direct reaction to the perceived dominance of western culture.

For Futures, which itself has a largely western origin, the lack of cultural diversity has already been identified as an issue. "The future has been colonized. It is already an occupied territory whose liberation is the most pressing challenge for peoples of the non-West if they are to inherit a future made in their own likeness....The future is little more than the transformation of society by new western technologies" (Sardar 1997) On the other hand we should not forget the idea that many Futurists and others have propounded for some time, in a number of different forms, the rise of Asia and the Pacific Rim. The financial troubles of the late 90s have raised questions about the belief that as the "Asian Tigers" developed the focus of world economic activity would shift from the Atlantic to the Pacific, but there are other reasons to think that this might be only a postponement and not a cancellation. Whether this anticipated shift is related to the 50-60 year Kondratiev wave, which according to some commentators has seen locus of innovation move in subsequent waves from the United Kingdom, to Germany, the United States and Japan (Hall 1981) or the end of a 500 year cycle of western dominance based on the enlightenment in Europe (Robertson 1998), is less important than the consequences of it occuring. Such a shift could lead to the replacement of many western based beliefs and attitudes by those based in oriental culture. Why should we, in the west, expect the emergence of China or India to be based on our cultural assumptions, when they were only imposed on top of much older and deeper cultures by western colonization? Could the millennium celebrations then, ironically, be the wake for western civilization?

For Futurists the millennium provides considerable opportunities to benefit from the increase in interest in the future, but it also has potential difficulties. The history of Futures suggests that interest waxes and wanes, and the obvious possibility is a big build up in '99 followed by a hangover in '00. The chances are increased by the very nature of much of the apparent Futures output we are likely to see during 1999 and early 2000. Much of it will be good media but with little real Futures content. It is likely to be over-reliant on "gee-whiz techie" stuff, the more sensational the better. The treatment of the genetically modified foods debate in the United Kingdom in early 1999 is a warning; with emotive headlines like, "Frankenstein Foods", but little real debate outside the serious television programs and broadsheet newspapers. There are likely to be many eye-catching predictions. It is evident from the requests made by journalists that this is what they are looking for and it is tempting, in order to gain publicity for Futures work, to provide them, but they have a clear potential to become hostages to fortune or subject to ridicule.

Ready to provide the required copy are a wide range of millennia-lists many of whom regard the year 2000 as a watershed in human existence. The Center for Millennial Studies, http://www.mille.org, provides a useful starting point for the examination of these ideas. Not surprisingly, several Christian groups regard the millennium as either the apocalypse itself or the beginning of a series of events that will culminate in the Second Coming and the end of the world as we know it. In 1997, Richard Landes, Director of the Centre for Millennium Studies predicted the convergence of a large number of pilgrims on Jerusalem, many of whom, he suggested would not anticipate returning to their place of origin (Landes 1997). On January 10th, 1999, *The Observer* reported the deportation by the Israelis of 11 members of the Concerned Christians whose leader is reported to have claimed that he will die in Jerusalem in December and be resurrected three days later (Sharrock 1999). It is not only Christian groups, according to Landes, that have "fixated on 2000 or thereabout as a key apocalyptic date," but also some "Jews, Hindus, New Age "pagans", Muslims and even secular "scientists."

The more extreme predictions associated with the Millennium Bug have much in common with these ideas and some have suggested it will be the trigger for the apocalypse. Whatever the consequences of the Bug, it has potential difficulties for Futurists. It is a no-win situation. If the major disruption that some predict occurs, the accusation will be, "Why did you not warn us?", on the other hand, if it turns out to be the damp squib that others anticipate, the comment will be, "What was all that fuss about?" Whichever happens we may expect questions about the value of Futures activity to be raised.

What then should Futurists do to maximize the benefit, but minimize the damage?

Firstly, the millennium does offer a once in several lifetimes opportunity to place the need for discussion of possible, probable and preferable futures in the spotlight. It is an opportunity we cannot afford to miss, but we need to use it carefully and accept that there will be a great deal of other future orientated activity that we would not regard as Futures but may get confused with it. Some of this material may, in fact, be usable by Futurists either as a basis to raise issues for debate or as a learning device to point out the uncertainties of prediction and the need for a more considered approach to Futures.

Secondly, we should be prepared for the possibility of a reaction after the event, when many predictions concerning the millennium will not come true. Which ones that will be will only become clear after the millennium, although most of us will have our own beliefs beforehand as to which it will be. The very extent of the predictions that will be generated around the millennium and the 21st century will, as many of them do not come true, provide a great deal of ammunition for those who argue that if as Futurists we cannot foretell the future we are no better than soothsayers. It would be wise, therefore, to rehearse our most convincing arguments for Futures in the knowledge, that to some they will always appear to be rationalizations of an untenable position. Here again, some of the forecasts may offer some help, as not all will have been made with the intention of being fulfilled; some may have been made as warnings to encourage changes in behavior that would invalidate them; others for clearly propaganda purposes. An investigation of the assumptions behind such forecasts and if evident the methods used to make them could also be turned to effective use to encourage a more critical approach.

Thirdly, we should use the millennium as an opportunity to broaden our focus from an essentially uni-cultural approach to a truly multi-cultural one. The twentieth century has been dominated by the West and its attitudes, values and interests, but there is enough evidence already to hand to suggest that to assume that this is sustainable in the third millennium would be unwise. In our increasingly global society Futurists can play an important role in helping us deal with the need to live with alternative pasts and presents, or at least differing perspectives on them, as well as futures.

Fourthly, it will be an opportunity to re-examine our own activities and assumptions. There is now a considerable stock of Futures work, accumulated over half a century, which we might usefully review. From it we might be able to assess our own claims about our work; understand the developments that it has gone through, which have hopefully led to improvements as the result of experience; and embark on a new millennium with a clearer idea of the contribution that as Futurists we can make.

REFERENCES

1. Clarke, T and S. Clegg. *Changing Paradigms: The Transformation of Management Knowledge for the 21st Century.* London: Harper Collins Business, 1998.
2. Coates, J F, Mahaffie J B and A. Hines. *2025: Scenarios of the US and Global Society reshaped by science and technology.* Greensboro: Oakhill Press, 1997.
3. Dator, J (editor). "Futures Studies in Higher Education American Behavioral Scientist," 42(3), *Sage Publications*, Thousand Oaks, 1998.
4. Galt, M., G. Chicoine-Piper, N. Chicoine-Piper, and A. Hodgson. "Idon Scenario Thinking: How to Navigate the Uncertainties of Unknown Futures Idon," *Pitlochry*, 1997.
5. Gardine, J. "Future Shock," *Bizarre*, March 18, 1999.
6. Hall, P. (1981) "The Geography of the Fifth Knodratieff Cycle," *New Society*, 55(958), 535-537, 1981.
7. Landes, R. Prospects for the Impact of Christian Apocalyptic Expectation on Israeli Politics and Society http://www.mille.org/Policy-Jews.htm, 29/01/99, 1997.
8. May G. H. "The Sisyphus Factor or a Learning Approach to the Future," *Futures*, 29(3), 229-242.
9. Naisbitt, J. *Global Paradox*, William Morrow, New York, 1994.
10. Northcott, J. *Britain in 2010*, London: Policy Studies Institute, 1991.
11. Office of Science and Technology. *The Future in Focus: A Summary of National Foresight Programs*, London: Department of Trade and Industry, 1998.
12. Ringland, G. *Scenario Planning: Managing for the Future*, Chichester, 1998.
13. Robertson, J. Beyond the Dependency Culture, A Talk to the UK Futures Group, London, September 19, 1998.
14. Sardar, Z. The problem, Seminar, New Dehli, 460, A Symposium on Critiques of Dominant Futures Studies, pages 12-18. To be published in *Rescuing all our Futures*, London: Adamantine, 1997
15. Schwartz, P. *The Art of the Long View: Scenario Planning— Protecting your Company against an Uncertain World*, Century Business, London 1991.
16. Sharrock, D. (1999) Swift exodus for 'pilgrims,' *The Observer*, p. 22, London, January 10, 1999.
17. Slaughter, R. "Looking for the Real Megatrends," *Futures*, 25(8) 827-849, 1993.
18. Van Der Heijden, K. *Scenarios: The Art of Strategic Conversation*, Wiley, Chichester, 1996.
19. White, R. "Futurologists Aren't what they Used To Be, *The Sunday Times*, News Review p. 5, London, January 21, 1999.

APPENDIX 1 SOME 2000 STUDIES

1963
Hall, P. London 2000, Faber and Faber, London.
1967
Kahn, H. and A.J. Wiener. The Year 2000: A Framework for Specula-
tion on the Next Thirty-three Years, Macmillan, New York.
1969
Hoivik, T. (ed) The year 2000 Pax, Oslo (in Norwegian).
Jungk, R. and J. Galtung. (eds) Mankind 2000, Universitetsforlaget,
Oslo.
1977
Hall, P. (ed) Europe 2000, Duckworth, London.
1979
Interfutures. Facing the Future: Mastering the Probable and Manag-
ing the Unpredictable.
Organization for Economic Cooperation and Development, Paris.
1980
Barney, G. O. The Global 2000 Report to the President of the US:
Entering the 21st Century, Volume 1 The Summary Report, Pergam-
on, New York.
1983
Hook. K., I. Nilsson, and A. Newton SWadeskog in The Year 2000:
A Different Kind of Swedish Municipality, Jordens Vanner, Stock-
holm (in Swedish).
Hook, K., I. Nilsson, and A. Young Wadeskog in Norden in The Year
2000, Akademillitteratur, Stockholm (in Swedish).
Williams, R. Towards 2000, Chatto & Windus, London.
1984
Bestusjev-Lada, I. The World in The Year 2000—a Soviet Global
Prognosis, Politisk Revy, Copenhagen, (in Danish, original in
Russian).
1987
Hompland, A. (ed) Scenarios 2000: Three Images of Norway,
Universitetsforlaget, Oslo (in Norwegian).
UNESCO Bureau of Studies and Programming, The World by The
Year 2000: Studies and Documents on Major Program I: Reflections
on World Problems and Future-Oriented studies UNESCO, Paris.
1988
Strandbakken, P. Sheltered Workplaces Towards The Year 2000,
Arbeidsmarkedsbedriftene, (in Norwegian).
1990
Kinsman, F. Millennium: Towards Tomorrow's Society, W.H. Allen,
London.
Naisbitt, J. and P. Aburdene. Megatrends 2000: The Next Ten Years...
Major Changes in Your Life and World, Pan Books, London.
1991
European Commission Europe 2000: Outlook for the development of

232

the Community's territory, Office of the Official Publications of the European Communities, Luxembourg.
Garett, M. J., G.O. Barney, J.M. Hommel, and K.R. Barney. (eds) Studies for the 21st Century, Institute for 21st Century Studies, UNESCO.
lists the following countries 2000 studies: Ecuador, Iceland, Norway, Ireland, Poland, Turkey, Mauritius, India, Taiwan, Japan

CREATING AND SUSTAINING SECOND-GENERATION INSTITUTIONS OF FORESIGHT

by

Richard A. Slaughter

The Commission for the Future was launched in a blaze of publicity in early 1986. It existed in one form or another for 12 years, had four directors, spent in excess of AUD $8 million, was privatized and vanished from public view during 1996. After many ups and downs, after a number of false dawns and unsuccessful attempts at revival, the last chairman of the board ran out of inspiration in June 1998 and the (CFF) closed its doors for the last time.

Over this period many people suggested that the CFF was never a fully satisfactory organization and, to a considerable extent, I concur. However, it seems to me that the conclusions that have been widely drawn are wrong: 'we've been there, done that; it did't work, so the whole idea of an organization focused specifically on the future should be abandoned.' It seems to me that if this notion persists then we will certainly be in for a much rougher ride in the early 21st Century than anyone would rationally desire. So the purpose of this paper is to suggest that, far from dismissing them, *we can learn from and apply the experience of first-generation IOFs* like the CFF. An international program of study and research is urgently needed for this purpose. If we wish to exert any real control, claim any sort of autonomy over our future, intending social innovators will deliberately embody these institutional learnings in a whole new generation of Institutions of Foresight.

WHY ARE INSTITUTIONS OF FORESIGHT NEEDED?

To begin with, it is patently clear that whether our concern is with our families, a business, a country or the future of the whole global system, at each of these levels we face unprecedented challenges from what Jim Dator calls 'tsunamis of change'. While any one change process can readily be exaggerated, over-hyped, it should be obvious to anyone who cares to look that the on-rushing waves of social, economic, technical and environmental change that we confront, together make up an outlook which is novel in the history of our species. That is why I refer to it as the 'civilizational challenge'.[1] It seems to me that what motivates most futurists—and certainly those who have devoted their lives to this area—is a sense that we should, as individuals, organizations and as a species, learn to pay attention. That is, to read the signals of change and act

---

Richard A Slaughter *is an independent futurist and director of the Futures Study Centre in Melbourne, Australia.*

accordingly. But this is asking for something that goes a long way beyond traditional expressions of prudence and foresight that can be found in various cultures.[2]

As is well known, short-term thinking rules in governments, education systems and, with some exceptions, in business too. I regard this as one of the main 'perceptual defects' that we have collectively inherited from the industrial era. It can be called such because it actively de-focuses and de-emphasizes the very innovative process that constitutes a historical breakthrough and which is comprehensively needed in our time. The breakthrough I am referring to is *a well-grounded and coherent forward view*. Short-term thinking thus pushes out of sight the source and springboard for rationales and strategies of adaptive change. That this is not merely an oversight can be seen when we consider aspects of dominant ideologies. For example, in his trenchant critique of corporatism, John Saul has this to say. He writes:

> corporatism—with its market—and technology-led delusions—is profoundly tied to a mechanistic view of the human race. *This is not an ideology with any interest in or commitment to the shape of society or the individual as citizen.* It is fixed upon a rush to use machinery—inanimate or human—while these are still at full value; before they suffer any depreciation.[3] (My emphasis.)

This passage helps to explain why, in a broader social sense, there is so little structural support for long-term thinking. While there are a number of government-driven 'foresight' initiatives in several nations, these are recent developments, the outcomes of which remain uncertain. Most forward-looking initiatives remain associated with technology trends, conventional (short term) planning, commercial or financial speculation and the development of corporate strategies. To date, the amount of futures work carried out by public bodies in the public interest is minimal. This is a huge oversight. The forward view is too important, too central to developing high-quality responses, to be marginalized. Yet that remains the current state of play.

The Australian Commission for the Future was one of a number of national government supported foresight initiatives created during the 70s and 80s. As with the premature closure of the American Office of Technology Assessment (OTA), its demise is to be regretted. But, properly understood (ie. in terms of a process of social and cultural legitimation), IOFs are experimental organizations created to address new needs and to explore strategies of a type and scale that are historically unprecedented. The tradition that they draw upon (the emergence of Future Studies as a metadiscipline) is itself only three or four decades old. So it is entirely understandable that some of them will fail. But the closure of any single IOF—or even of a number of them—should not be taken to indicate that societies do not need well-grounded foresight. In fact the opposite is the case. They provide a range of vital services for societies undergoing the stress of rapid structural change.

WHAT SERVICES DO INSTITUTIONS OF FORESIGHT PROVIDE?

A few years ago I surveyed a sample of internationally-significant IOFs and derived an overview of their activities. Here, in summary, is an overview of the services they provide. First, it is clear that they *raise issues of common concern* that are overlooked in the conventional short-term view; e.g. issues about peace, environmental stability, inter-generational ethics, the implications of new, and expected, innovations, both social and technical. Second, they *open out the forward view* and, in so doing, highlight dangers, alternatives and choices that need to be considered before they become urgent. Third, they *publicize the emerging picture of the near-term future* in order to involve the public and contribute to present-day decision-making. Fourth, they *contribute to a body of knowledge about foresight implementation* and the macro-processes of continuity and change that frame the future. Fifth, they *identify some of the dynamics and policy implications of the transition to sustainability*. Sixth, they help to *identify aspects of a new world order* so as to place these on the global political agenda. Seventh, they *facilitate the development and application of social innovations*. *Eighth, they help people to deal with fears and become genuinely empowered* to participate in creating the future. Ninth, they *help organizations to evolve in appropriate ways. Finally, they provide institutional shelters* for innovative people and for experimental, or public interest, futures work which, perhaps, could not easily be carried out elsewhere.[4]

It should be obvious that such contributions help in many practical ways to initiate and support the crucial shifts of perception, policy and practice which, in no small way, form the pivot upon which our over-heated and over-extended global "megaculture" now turns.

RISE AND FALL OF THE CFF

As a pioneering Institution of Foresight (IOF) the CFF was, throughout its life, under-equipped and under-designed. It attempted to carry out a wide range of projects and initiatives, many of which were intended to raise public awareness. But projects were usually issues-based and it was not until rather late in the piece that standard futures methodologies (to enable more sophisticated options) were even contemplated. In this it differed from many other IOFs. The initial selection of staff was dictated more by a political agenda than a professional one, and this colored the nature of the organization from the start. It is startling to realize that at no time thereafter did any full-time employee possess a background in Future Studies. To be fair, qualified futurists were not, and are still not, very numerous. But neither were steps taken to ensure that key staff acquired the necessary grounding. So the CFF was flawed from the very beginning. Hence it is reasonable to conclude that the *impulse* underlying its creation was well-founded, but the execution failed for a number of reasons. These include the following.

- Lack of comparative knowledge from other foresight contexts.

- Role conflicts that arose from being dependent upon a government department (Science); whereas futures work arises from a wider context and may involve challenging political priorities.

- The specialized knowledge, concepts, methodologies etc. available in Future Studies were prematurely dismissed (in favor of consciousness-raising). These resources could have helped the CFF to develop in more productive and professional ways.

- Lack of clarity about the purposes and practices of futures work created a policy vacuum which made it difficult to adjudicate the many claims made upon the CFF and especially its director(s).

- There was a consistent failure to employ suitably qualified personnel with a background in Futures Studies.[5]

Against the early criticisms were a number of successes. The *Greenhouse Project* was probably the most successful of all CFF initiatives in that it helped make the concept a household term in Australia. *The Bicentennial Futures Education Project* (BFEP) helped to place futures studies on the educational map and produced some useful materials.[6] The CFF also produced a range of other publications, some of reasonable, or better, quality. Also, an intense schedule of public briefings, radio shows, parliamentary seminars and the like certainly helped to influence public understanding and raised the profile of concerns such as innovation, re-cycling and the meaning of 'the clever country'.

The glossy journal *21C* was the most high-profile of the CFFs publications. Its own rise and fall parallels that of the CFF in some ways.[6] It was launched in 1990 as an over-designed, large-format, magazine with the sub-title 'Previews of a Changing World' and a fairly standard menu of futures-related articles. Several years later it was taken over by a commercial publisher who, in 1998, pulled the plug due, it is said, to lack of advertising. Over eight years and 26 issues it had evolved from the simplistic 'previews' format to a highly specialized one catering to a limited, but discriminating, readership. The subject matter had changed. *21C* had become an ultra-sophisticated cultural studies journal with a focus on 'the impact of technology on culture'. The design standards were exemplary; but to my mind it had forsaken the futures arena and wandered too far into the detailed exploration of what I can only call 'the detritus of post-modernism': the realm of post-modern gurus, technological breakthroughs, media and, especially, the world of the internet. The whole process shows what can happen in the absence of sound editorial guidelines based upon a coherent disciplinary foundation. However *21C* was certainly one of the most (if not *the* most) exciting and original publications ever produced in Australia and its demise left a significant gap.

Meanwhile the CFF had continued on its own long, meandering journey via the regimes initiated by four very different directors. When the last resigned in 1996 there was a significant hiatus while the board considered its options. The position of director was advertised, but a director was never appointed. A *rapprochement* with Monash University, in Melbourne, was pursued, but failed. Finally, an office minder was hired on a part-time basis and at that time the CFF ceased to be a viable entity. By late 1997 the website had been virtually abandoned. Perhaps the last creative gasp was the belated attempt in 1998 to launch a *Future Directions* newsletter. But the modest 8-page format was unoriginal and unproven; the nearly AUD $200 annual fee too high to attract sufficient subscribers. *Future Directions* expired in June 1998 after only 3 issues.

Looking back over the 10-12 years of its existence, I am firmly of the view that the CFF was by no means a waste of time and money. Rather, it encountered forces that it was ill-equipped to face, let alone resolve. The fact that such institutions are rare, are not widely supported, and are certainly not widely understood, points beyond the analysis of particular IOFs to the social context in which they are embedded. Perhaps the experience of the CFF reveals something about the 'shadow side' of human societies as we head into the millennium period.

ACKNOWLEDGING 'THE SHADOW' IN HUMAN SOCIETIES

It is easy to focus a critique of first-generation IOFs such as Australia's CFF on weaknesses in the original design, deficiencies in the way it was administered and led and the half-hearted nature of its 'stop/start' work program. There is, as I have noted, some truth in all of these. But, in the context outlined above, this 'internal' diagnosis is both unconvincing and insufficient. I therefore tend toward an explanation of a rather different order: one that allows us to build on the mistakes and the successes of organizations like the

CFF, and to move forward. This leads toward a powerful and disturbing conclusion.

On the bright, superficial side of human experience, most people are keenly aware of the way that powerful new technologies are being promoted with the promise that they will support millions of people in unprecedented wealth and comfort. But, at a deeper level, I don't think that many really believe it—least of all the young. If we look deep within 'the shadow' (i.e. the repressed contents of the human mind, both individual and collective) we find familiar defence mechanisms: avoidance, denial, lack of interest which, given the global context, clearly imply Dystopian 'breakdown' futures. Though the latter are highly plausible, and well-founded in known facts, they remain anathema to dominant institutions and the mass media and are thus ignored (except in entertainment where the rehearsal of disaster is a familiar theme, and one readily dismissed). However, leading practitioners within FS continue to suggest that the future of civilization now hangs between two worlds or, more appropriately, two kinds of world.' One is where the balance swings away from foresight and we learn (if we learn at all) through the kinds of social experience seen in the collapse of other civilizations, though on an immeasurably wider scale. The other is where humankind negotiates the end of the industrial period with foresight, elegance and skill and finds new ways to live on this over-stressed planet. In this latter world the forward view is a functional necessity, not an esoteric abstraction.

So could it be that the accusing finger should point beyond particular attempts at institutional innovation to the heart of our societies and major institutions? If so, does it not point to the fundamental assumptions, the views of reality, that still govern them? Notions of growth, of a powerful but defective economics, the primacy of the marketing imperative, the view of nature as a mere resource, of materialism, of the future as 'an empty space'—all these are powerful aspects of an existing worldview—though their 'use by' dates have, in many cases, long expired. It is within this arena of ill-considered but deeply-embedded cultural commitments and presuppositions that we may find the most profound explanation about why IOFs in recent years have had an up-hill battle, and why some of them no longer exist. In Australia, as in so many other places, forward thinking is neither a political habit, a widespread commercial practice nor a popular pastime.

But if IOFs became a more effective social force, it could become all of the above, and more.

DESIGN PRINCIPLES FOR SECOND-GENERATION IOFS

First-generation IOFs were created largely in isolation from each other and in the absence of a body of applied knowledge about how best foresight work in the public interest can, or should be, carried out. So, as noted, many of them failed. But, at the dawn of the Third Millennium, there are enough case studies, enough accumulated experience, to begin to derive some provisional lessons, or design principles, for future IOFs. The challenge is to assemble a body of applied knowledge that will form a sturdy foundation for 21st century IOFs.

One starting point is the outline research agenda for IOFs that Martha Garrett and I suggested in 1995.[8] Earlier, still-relevant, sources are the materials gathered by Clem Bezold and his associates from the use of state government foresight in the USA, and Lindsay Grant's book on *Foresight and National Decisions*.[9] There are also occasional more general overview-type studies such as that carried out by Skumanich and Silbernagel in 1997. These researchers studied what they termed seven 'best-in-kind' foresight programs and concluded that the most successful ones had the following features:

1. They began with a perceived need to prepare for future challenges.
2. They each had 'program champions' during the start-up period.
3. They proved responsive to client needs.
4. They involved the relevant participants in the process.
5. They experienced a legitimizing process.[10]

These are useful insights. It is the last factor which, I suspect, weighed heavily against the CFF. It was widely seen as a politically-driven entity, rather a commercial or professional one. It won few friends in the parliament, in business, in education, in intellectual circles or in contemporary social movements. Thus for most of its life it lurched from one crisis to another, despite the best efforts of the board and successive directors.

So where do we go from here? The following suggestions arise in the context of the CFF and a number of other examples.[11] It is essential that they are critiqued, checked and supplemented by further work. Nevertheless, they provide some clear starting points for enquiry and practice in this still under-developed, but increasingly vital realm.

1. DEFINE CORE PURPOSES

The core purposes of any IOF should be carefully defined and linked with the main institutional functions (as in a successful business). In other words, there must be a clear match between purposes and the structures created to sustain them. For example, there is little merit in creating a 'soft', inspirational, consciousness-raising operation if the main clients are likely to be results-oriented government or business people. Ends and means must be appropriate to the chosen purposes.

2. FUNDING

Funding issues should be tackled and a secure, diversified basis of financial support established as soon as possible. During the early years any IOF will be vulnerable to many hazards, not least of which is running out of funds. Two starting strategies are as follows. One is to secure benefactor funding. This means locating an organization or an individual who will elect to support the new organization because they believe it is worth doing, i.e., that there are intrinsic benefits, or tangible benefits, or both. Such support is not impossible to find if it is sought in the right places. A second approach is to establish a fee for service operation from the outset. If successful, this becomes a source of 'hard' money that will not suddenly disappear. By 'fee for service' I mean a viable product or service that is offered for sale. It may be a series of foresight-related courses or seminars, a new angle on consulting or a publication such as the Worldwatch Institute's very successful *State of the World* series.

3. CONTEXTUAL KNOWLEDGE

The knowledge gained from other foresight initiatives around the world should be thoroughly understood and applied such that the learning curve can begin from a higher level and take place more quickly. There is nothing more futile than for different foresight initiatives to be each re-inventing similar 'wheels', as it were. So to make the very best use of scarce resources and personnel, every effort should be made to learn about the nature, work and strategies involved in other IOF initiatives. In part this means that the channels of communication between widely-dispersed IOFs should be open and facilitative (see below.)

4. QUALITY CONTROL

This is of such overwhelming importance that I would recommend it be taken as a central principle of any IOF. The reason is that there are many myths and misunderstandings about FS and futures work in general. Second rate futures work is worse than none at all because it provides spurious grounds for the dismissal of the whole enterprise. Measures for quality control include: external refereeing, benchmarking, best practice criteria and the adoption of a code of professional ethics.[12]

5. QUALIFIED EMPLOYEES

Those working in IOFs should be fully qualified to carry out futures work. This will necessarily mean that a certain proportion of employees will either have recent relevant experience of futures-related work or will undertake the necessary training as a condition of their employment contract. While there may be regrettable consequences attending the professionalization of FS work, these are minor compared with the need to create durable organizations with professional standards.[13]

6. USE OF ROBUST METHODS

Futures methodologies have developed rapidly in the last decade or so. The field is no longer limited to the earlier empirical methods (eg. forecasting, trend extrapolation, scenarios) that were developed in the 60s and 70s. These will remain important in relation to the empirical dimensions of futures problems. But more sophisticated work will also integrate other approaches. Some years ago Inayatullah suggested a three-fold division into empirical, critical and interpretative approaches. [14] This represents a useful starting point for the wider exploration and use of futures methods which go a long way beyond the pop-futurist habit of merely re-hashing surface understandings which are normally both highly problematic and culture-bound.[15]

7. CONSTITUENCIES OF SUPPORT

Particular attention must be paid to building up and sustaining the constituencies upon which such enterprises depend. This is a challenging task since the spread of interests is clearly very wide. In this regard, full and proper use should be made of all available media outlets to ensure that they are informed in good time of all initiatives, publications etc. Key figures in relevant areas should be consulted and valued. Obviously all of the above takes time. Yet so vital is this area that the appointment of a full-time PR individual need not be seen as a luxury. It goes without saying that all IOFs require a board comprised of leading people from key social and economic areas who will tenaciously pursue the interests of the organization. Such work embraces basic fund-raising as well as the search for social legitimation (see below).

8. COMMUNICATION

One of the greatest lacks at the present time is a dedicated channel of communication for IOFs and those who work in them. Of course, all are loosely connected by the internet, journals, futures organizations and the like. But it may be that the time is right to establish a more formal association to assemble foresight knowledge and expertise in a more coherent and reliable form. One approach would be to establish a dedicated web site. Another would be to host special-interest gatherings in collaboration with, e.g., WFS and WFSF conferences. Overall, it is vital that IOFs communicate with similar organizations around the world and begin to: share expertise, organize meetings, pool efforts in common projects and, perhaps, to begin to 'speak with one voice' across cultures and national boundaries. IOFs that begin to cooperate in these ways will be able to wield far greater social and symbolic power than any of them could working alone.

9. LEGITIMAZATION

The parent field of FS is arguably making steady progress in its own path toward full social and professional legitimazation. That is, it is emerging as a serious and substantive entity that can contribute in a host of ways to the framing of policy and practice in many, many fields.[16] If, as a group of organizations dedicated to some of the ends outlined here, IOFs are able to set themselves appropriate tasks to serve their constituencies in a competent, consistent and high-quality way, then they too will follow this path. It will then become self-fulfilling: IOFs will finally have 'arrived' in the sense of having established their social, economic and professional viability. They will be seen as legitimate, socially-vital organizations called forth by the historical conditions of our time and serving a range of profound human interests.

10. RESEARCH

As part of the coming-together of IOFs around the world, there is a need for an intensive program of work to carry out tasks such as the following:

- To document as much institutional foresight activity as possible. This will form the history of the discipline of foresight work. It will record the emergence of this sub-field and its early attempts to get established.

- To investigate particular IOF case studies in order to draw out, check, critique, create and re-create the essential procedures and principles of operation that minimize failure and maximize the chances of success in particular contexts.

- To explore ways of functioning, modes of cooperation and the like that will increase the efficiency and effectiveness of IOF work around the world.

- To evaluate the outcomes of IOF activity in relation

to appropriate professional standards.

In such ways IOFs can be created and sustained into the distant future.

CONCLUSION

From IOFs to National Foresight Strategies

The creation of free-standing IOFs is itself only part of a wider foresight strategy; an attempt to help societies move from their traditional stance of being inspired or 'driven' by the past, to a stance which remains open to the past but which is also increasingly responsive to the 'signals' of the near-term future, as revealed by well-constructed forward views. To achieve this goal each nation will find it valuable to undertake a national integrated foresight strategy that matches its own priorities, needs and culture(s).

In a paper published in 1996 I outlined a rationale and approach to the creation of a national foresight strategy for Australia. The basic steps were as follows:

1. Create an Australian Foresight Institute (AFI).
2. Map national and international foresight work.
3. Develop a skill-transfer strategy.
4. Identify key sectors, organizations and individuals within each.
5. Review progress and link with similar initiatives elsewhere.
6. Secure long-term funding.[17]

Three years later a feasibility study for the AFI was completed and accepted by a leading university in Melbourne. The new organization began work in mid-1999. This development will help to catalyze the other steps mentioned above. It is a long-overdue development, though not a new idea. National 21st century studies have been around for some time (although, puzzlingly, Australia did not participate in them).[18] Nevertheless, support for such work has come from a highly significant source. In an article about a submission to the Australian government's 'West Review of Higher Education', Prof. Don Aitkin, vice-chancellor of the University of Canberra had this to say. He wrote:

> It seems to me that humanity may have only two gen-
> erations left in which to sort out how to modify the
> impact of the human species on the planet. If it does
> not learn how to do that, then the world is likely to

experience a catastrophe even more severe than that which followed the collapse of the Roman Empire. Compared with 1500 years ago, we do know in some detail what is happening and we know at last some of what needs to be done. Moreover, we understand that where we do not know something, we can set about finding it out.

He then added:

The principal institution in humanity's race to save itself, if we set aside enlightened governments, is the modern university.[19]

Universities are not the only places where IOFs can be created and sustained. But such developments are clearly appropriate there. Indeed, properly understood, universities can be considered IOFs in their own right.[20] Such developments would go a long way toward enabling national foresight strategies appropriate to the needs of different societies and cultures.[21]

In time, we can expect to see a variety of second-generation IOFs springing up internationally. They will pursue a variety of agendas and serve a variety of interests. There is no single mould: they will be multicultural and diverse. Within that diversity will lie part of their strength. The latter will also derive from rigorously interrogating worldview assumptions, dealing effectively with the many avoidances and repressions that are expressions of 'the shadow', as well as using and adapting some of the more practical suggestions outlined here.

By or before 2010 we can anticipate an integrated network of IOFs operating around the world to create, sustain and apply the forward view in a wide variety of contexts. These purpose-built institutions will be durable, far-sighted, but keenly attuned to their social contexts—and thus socially valued. They will help to plot dynamic and sustainable paths ahead in the manifestly challenging and unstable conditions of the early 21st Century.

This is an historic task wherein there is nothing to lose and just about everything to gain.

NOTES

1. R. Slaughter. *The Foresight Principle: Cultural Recovery in the 21st Century*. Praeger, Westport, Connecticut, 1996.

2. G. Mander's *In the Absence of the Sacred*. Sierra Club, San Francisco, 1991, contains a number of examples of practical foresight measures from traditional Indian cultures.

3. J.F. Saul. *The Unconscious Civilization*. Penguin, Melbourne, 1997,

pp. 162.

4. R. Slaughter 1996 op cit note 1, pp. 106-107.

5. Readers wishing to review an fuller account of the first 6 years of the CFF may wish to see: R. Slaughter, Australia's Commission for the Future: the first six years, *Futures*, 24, 2, April 1992, pp 268-276.

6. *21C* was originally published by the Commission for the Future in 1990. It was purchased by publishers Gordon and Breach in 1993/94 and closed in 1998 after 26 issues.

7. H. Tibbs. "Global Scenarios for the Millennium," *ABN Report*, Vol 6 No 6, 1998, pp. 8-13 Prospect, Sydney.

8. R. Slaughter and M. Garrett. "Towards an agenda for institutions of foresight," *Futures*, Vol 27 No 1, January 1995, pp 91-95.

9. See L. Grant. (ed) *Foresight and National Decisions*. Univ. Press of America, Lanham, MD, 1988. Also C. Bezold, Lessons from State and Local Government, in Grant, L. (ed) Ibid 1988, pp 83-98.

10. M. Skumanich and M. Silbernagel. *Foresight Around the World: A Review of 7 Best-In-Kind Programs*. Battelle Seattle Research Centre, 1997.

11. A sample of 7 IOFs are considered in R. Slaughter 1996 op cit note 1, chapter 7. Other examples can be located via eg. the *Futures Research Directory: Organizations and Periodicals*, 1993-4, World Future Society, Bethesda, MD, 1993 and the World Futures Studies Federation's annual membership directory.

12. See Bell, W. A. "Futurist Code of Ethics," in R. Slaughter (ed) *The Knowledge Base of Futures Studies Vol 3: Directions and Outlooks*. Futures Study Centre, Melbourne, 1996, pp 97-111.

13. R. Slaughter. "Professional Standards in Futures Work," forthcoming, *Futures*, Vol. 31 No. 1, 1999.

14. S. Inayatullah. "Deconstructing and Reconstructing the Future," *Futures*, Vol. 22 No. 2, March 1990, pp. 115-141.

15. For a recent example of this type of work see G. Glenn and T. Gordon. *1998 State of the Future*, American Council for the United Nations University, 1998; also review of same: R. Slaughter, Projecting the Millennium, *Futures*, Vol. 31 No. 1, 1999 (forthcoming).

16. J. Dator. (ed) Special issue of *American Behavioural Scientist*, on Futures Studies, Vol. 42, No. 3, November/December 1998.

17. R. Slaughter. "A National Foresight Strategy for Australia," *ABN Report*, Vol. 4 No. 10, 1996, pp. 7-13.

18. See M. Garrett. (ed), *Studies for the 21st Century*. UNESCO, Paris, 1991.

19. D. Aitkin. University Challenge, *Australian Higher Education Review*. Weds. March 19th, 1997, p. 33.

20. R. Slaughter, Universities as Institutions of Foresight, *Journal of Futures Studies*, Vol. 3 No. 1, 1998, pp. 51-71, Taiwan, Tamkang University.

21. See P. Wildman and S. Inayatullah. *Futures Studies: Methods, Emerging Issues and Civilizational Issues*, (CD-ROM), Prosperity Press, Brisbane, 1998.

THE PROFESSIONAL
MEMBERSHIP PROGRAM
of the
WORLD FUTURE SOCIETY

Since January 1985, the World Future Society has offered a special program for members who are involved professionally in futures research, forecasting, corporate or institutional planning, issues management, technology assessment, policy analysis, urban and regional planning, and related areas.

Professional Membership includes:
- All benefits of regular membership in the World Future Society, including a subscription to *The Futurist*.
- A subscription to *Futures Research Quarterly*, the Society's professional journal.
- Society supplement newspaper *FutureTimes*.
- Invitations to meetings arranged specifically for professional members. The registration fee for these occasions is very modest. Recent forums were held in Atlanta, Georgia, Washington, D.C., San Francisco, California, Chicago, Illinois.
- A copy of the *Futures Research Directory: Individuals*. Published in 1995, to be updated in 1999, this comprehensively-indexed, 235-page volume lists nearly 1,000 futurists.

Dues: The dues for Professional Membership are $105 per year.

Comprehensive Professional Membership includes all benefits of Professional membership plus a subscription to *Future Survey* and complimentary copies of all other books published by the Society during the term of membership.

Dues: The dues for Comprehensive Professional Membership are $195 per year.

These programs are available for individuals only. To join, or for information about Institutional Membership, contact Society headquarters:

World Future Society
7910 Woodmont Avenue
Suite 450
Bethesda, Maryland 20814
USA

Or call 301/656-8274
or fax 301/951-0394

TENTATIVE SCHEDULE OF UPCOMING
WORLD FUTURE SOCIETY MEETINGS

July 23 - 25, 2000
2000 Annual Conference
Future Focus 2000: Changes, Challenges & Choices
Westin Galleria Hotel
Houston, Texas
July 26, 2000
Professional Members' Forum

July 29 - 31, 2001
2001 Annual Conference
Hilton Minneapolis and Towers
Minneapolis, MN
August 1, 2001
Professional Members' Forum

(For further information write: World Future Society, 7910 Woodmond Avenue,
Suite 450, Bethesda, MD 20814, or call 301/656-8274)